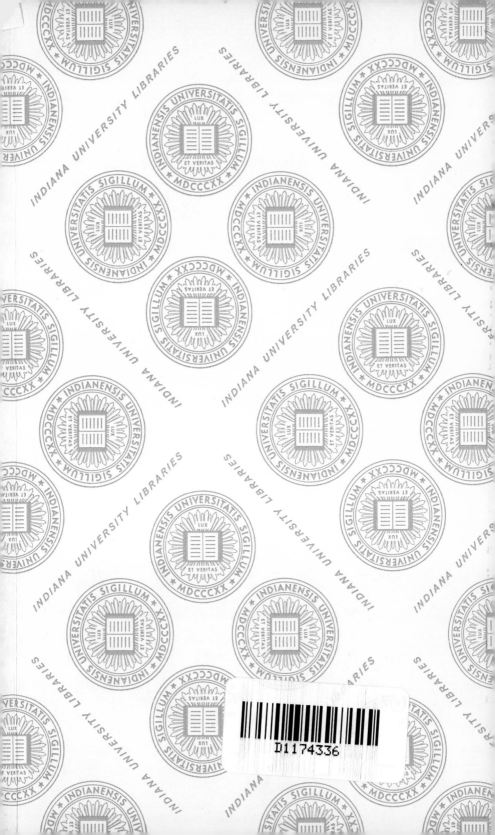

Biological Techniques Series
Alexander Hollaender, *Editor*

*Autoradiographic Techniques: Localization of Radioisotopes
in Biological Material*
WILLIAM D. GUDE

Introduction to Research in Ultraviolet Photobiology
JOHN JAGGER

The Laboratory Mouse: Selection and Management
M. L. SIMMONS AND J. O. BRICK

Introduction to Research With Continuous Cultures
H. E. KUBITSCHEK

A Manual of Basic Virological Techniques
GRACE C. ROVOZZO AND CARROLL N. BURKE

A Manual of

BASIC

VIROLOGICAL

TECHNIQUES

Grace C. Rovozzo

Eastern Connecticut State College
Willimantic, Connecticut

Carroll N. Burke

University of Connecticut
Storrs, Connecticut

Prentice-Hall, Inc., Englewood Cliffs, N.J.

Library of Congress Cataloging in Publication Data

ROVOZZO, GRACE C.
   A manual of basic virological techniques.

   (Prentice-Hall biological techniques series)
   Bibliography: p. 274
   1. Virology—Technique. I. Burke, Carroll N.,
joint author. II. Title.
[DNLM: 1. Virology—Laboratory manuals. QW 25
R875m 1973]
QR360.R63   576′.64′028   73-2020
ISBN 0-13-553289-2

PRENTICE-HALL INTERNATIONAL, INC., *London*
PRENTICE-HALL OF AUSTRALIA, PTY. LTD., *Sydney*
PRENTICE-HALL OF CANADA, LTD., *Toronto*
PRENTICE-HALL OF INDIA PRIVATE LIMITED, *New Delhi*
PRENTICE-HALL OF JAPAN, INC., *Tokyo*

© 1973 by Prentice-Hall, Inc., Englewood Cliffs, New Jersey

10 9 8 7 6 5 4 3 2 1

Printed in the United States of America

CONTENTS

THREE   Tissue Culture Propagation

FOUR   Propagation and Titration of Virus

FIVE   Serologic Techniques

SIX Biochemical and Biophysical Characterization of a Virus

SEVEN Histochemical Methods

EIGHT Bacteriophage

NINE   Preparation of Materials, Equipment, and Supplies

Appendix A

Appendix B

Bibliography   274

Index   279

PREFACE

This manual was prepared to assist those who must acquire the skills and become familiar with techniques commonly used in virology. Those who have used preliminary drafts of this book have come from many disciplines and various levels of experience. Early drafts were used by student assistants, many of whom were nonscience majors employed in our laboratories. Others who borrowed or extracted instructional sequences included biologically oriented graduate assistants, fellow faculty members, and post-professional associates who were D.V.M.s, as well as M.D.s and Ph.D.s.

Where translation facilities were lacking for certain visiting faculty from foreign universities or firms, the diagrams and photographs seemed to minimize the difficulties. With relatively little help, they could familiarize themselves with previously unfamiliar procedures by following the illustrations, with but brief reference to the text. Associates and graduate students have reduced to tatters the two painfully typed copies kept on hand as a reference source for laboratory personnel. We can only hope that it will be equally useful to others.

We wish to emphasize that this manual was designed to acquaint the user with procedures which are fundamental to the study of viruses. Obviously, there are methods which require great sophistication in instrumentation and experience. The procedures covered here should give the user a firm foundation upon which the more sophisticated studies can be based. In essence, the manual is designed for the student who is not yet in research and for the scientist who is no longer a student.

For their assistance is typing the manual, the authors wish to thank

Miss Ruth Carey, Mrs. Rose Cournoyer, Mrs. Diane Dessert, Mrs. Victoria Gordziel, and Miss Terry Williams.

GRACE C. ROVOZZO
Willimantic, Connecticut

CARROLL N. BURKE
Storrs, Connecticut

INTRODUCTION

Faculty will find this manual useful as a laboratory aid for students. Experiments may be set out via chapter or section. This provides all basic procedures which might be found in student manuals and the instructor may simply point out the appropriate section, i.e. . . . . "Next week you will prepare and inoculate chick kidney cell cultures. Read section *a* and *b* which we will do on Tuesday, and sections *c* and *d* for Thursday. All materials and methods to be used will be found there." Detailed explanations through-out all sections should enable the user to proceed with minimum supervision.

Post-professional staff and faculty will find the manual helpful for introductory research techniques. The researcher can select sections of pertinence without the handicap of utilizing materials and methods designed for student use, where equipment is usually supplied and sterilized, and the experience serves only as a brief excursion into the procedure.

Laboratory assistants will find formulas and techniques for preparing media, methods for assembly of glassware and other equipment, sterilization procedures, etc. without being concerned with test or culture procedures. Support for the written directions will be found via diagrams, tables, flow sheets, and photographs.

Procedures are designed around the use of viruses considered "non-pathogenic for humans". Obviously, it is strongly recommended that you familiarize yourself with a technique by using a "non-pathogen" before attempting any work with those viruses which require extreme care in their propagation and use.

A Manual of

BASIC VIROLOGICAL TECHNIQUES

Collection and Preservation
of Specimens
ONE

A. GENERAL CONSIDERATIONS IN THE
COLLECTION OF SPECIMENS

The experimental or field situation will suggest, in good part, the appropriate specimen and method of collection for virologic study. There are a number of factors, however, which will determine the effectiveness of the specimen collection. Among these are (1) time of collection, (2) tissue selected, and (3) the handling and storage of the specimens.

1. Time of Collection

It is best to obtain specimens from an animal as early as possible in the course of a disease. Avoid waiting to take specimens until an animal is moribund and near death. The site of multiplication (which is not always the site of response) may be so destroyed as to no longer yield viable virus. Whenever possible, specimens should be taken over a period of time, at least until the organ or system of viral multiplication is determined and a growth or multiplication study of the virus is done.

Another factor to consider is the increasing antibody levels in an infected animal. Full maturation and release of virus may be impaired under such circumstances and, again, taking samples over a period of time should be considered. Timed samplings are easily arranged under laboratory conditions but not as easily accomplished under instances of field outbreaks of diseases.

1

When death does occur, necropsy material should be collected immediately. Otherwise, viruses may die, bacteria may spread rapidly from the intestinal tract, and contamination of organs and tissues will be a major problem. It must be noted, however, that viruses have been successfully isolated from tissues of animals which have been dead for periods in excess of 24 hr. This is not the ideal and should be avoided where possible.

In the case of cell cultures, harvesting the cultures as soon as alterations are noticed is advisable. Otherwise the maturation and release of virus may result in such extensive cell death that free viruses, finding no other cells in which to multiply, themselves die off. If no changes are noted, harvesting (collecting) samples may still be in order. This may be followed by serial passages, serologic testing, or other procedures to determine the presence or absence of virus in the harvested cultures. Again, as in collecting specimens from living hosts, the harvesting of cell cultures at the proper time can affect the results of the tests which follow.

2. The Tissue Selected

The target organ or tissue of expression in virus infection may not be the only one to collect for virus isolation. Under field conditions, you are sometimes compelled to take samples without knowing the infectious agent or the organ in which it multiplied. This is also true when working with unknown viral agents under laboratory conditions. In either instance, a variety of specimens may be required in the preliminary stages of the study: heart, liver, kidney, lung, and fluids such as blood and spinal fluid, for example. In any instance, it is better to take more samples from the host or more frequent harvestings from infected cell cultures than to regret the absence of a vital specimen.

3. Handling of the Specimens

A. PRECAUTIONS

It is vitally important to exercise caution in handling animals and materials which are infected. Wear rubber gloves and use sterile procedures during necropsy. Handle living animals with due respect. Bites and scratches can be dangerous.

B. TRANSPORTATION OF SAMPLES

If samples are taken in the field, they should be taken to the laboratory in as short a time as possible. Shipping or carrying containers of insulating expanded plastic, ice-packed plastic bags, jars and tubes, or thermos bottles are commonly used. When dry ice is used for shipping, all containers must be tightly sealed since CO_2 can be detrimental to viruses.

C.   LABORATORY PROCESSING OF SAMPLES

Once the samples have been obtained, they will be processed (ground and suspended, diluted, etc.) in accordance with the purpose of the study. Details of the various methods are covered in Section B. Aside from the storage and testing of the samples by procedures common to virologic studies, certain support information should be obtained from additional studies, histologic examination of specimens under the light and (sometimes) the electron microscope, and the testing of the samples for bacteria. It is rare that any worthwhile virologic study would proceed without some of the specimen being processed for what might be termed *routine* histologic and bacteriologic checks. Unless you have experience in both these areas, it is best to consult those who have, in determining the tests most appropriate. In any instance, you will be required to submit your specimen in a form most convenient for the laboratory personnel involved. The procedures usually required for the histologic and bacteriologic examinations are given in detail in Sections C and D, respectively.

B.   SPECIMENS

1. Necropsy

a. Anesthetize the animal from which tissues are to be removed and clean the area of the intended incision with disinfectant. Alcohol (70%) is commonly used. With sterile instruments, remove the skin from a furry animal. This will reduce the chances of contamination from loosened animal hair. This is not usually necessary with smooth-skinned specimens, and cleaning the area of incision is usually sufficient.

b. With fresh, sterile instruments (not those used in removing the skin), grasp the organ or tissue to be removed; clip it loose from surrounding tissues; and transfer it to a sterile Petri dish, glass jar, or vial. Do not use preservatives. Affix labels.

c. Set aside sample for histologic examination (Section C).

d. Take preliminary test for bacteriologic examination (Section D).

e. Store specimens temporarily or process for long-term storage (Section E). It is preferable to process the samples for virologic isolation as soon as possible. Work with fresh unfrozen tissue is more apt to be successful than with frozen tissue since some cell-associated viruses may be killed or altered by freezing.

2. Making Tissue Suspensions

Tissues, including lung, spinal cord, kidney, etc., may be ground, minced, or made into a fine suspension by several methods.

There are several hand-operated grinders (homogenizers) of glass, or glass and Teflon, construction with which you should become familiar (Fig. 1-1). The Ten Broeck is designed for grinding small samples. Clearance between the chamber and the pestle will be 0.004 to 0.006 in. depending on the volume held by the grinder. The Dounce homogenizer is designed to reduce tissue to fine particle size without damage to cell nuclei. Interchangeable pestles permit large and small clearances. The Potter-Elvehjem-type grinder has a grinding chamber with 0.004 to 0.006 in. clearances. The Ten Broeck and Potter-Elvehjem types have pestles designed for insertion into the chuck of a stirring apparatus or for connection to a drive shaft.

Commercially available electrified homogenizers with variable speed controls from 100 to 45,000 RPM are also available. The cups and stainless steel blades can be autoclaved as a unit. These are ideal for preparation of highly pathogenic viruses where an aerosol-free assembly is desired and for preparation of embryo extracts with much higher yields.

For non-mechanical disruption of biological materials, ultrasonic waves may be used. Various commercial units are available which differ in their acoustical energy output, volumes to be handled, etc. Whatever system you use, observe proper sterile techniques throughout.

A. METHOD—GENERAL

Tissue disruption via mechanical homogenizers or ultrasonic waves will follow the same principles. The tissue must be cut into small pieces,

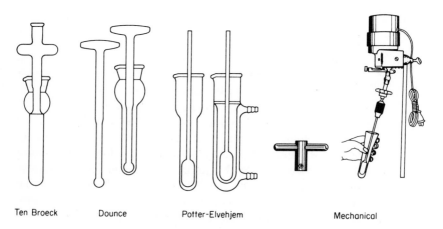

Ten Broeck   Dounce   Potter-Elvehjem   Mechanical

Fig. 1-1. Homogenizers.

added to the appropriate flask containing balanced salt solution and kept cold. Follow manufacturer's directions for operating the particular model being used. Complete the disruption of the tissue as described in the following sections.

B.   USING HAND-OPERATED HOMOGENIZERS
(TEN BROECK, ETC.)

(1) Unwrap the homogenizer, remove the mortar, and place it upright in a test tube rack.

(2) Remove the fresh tissue (or thawed sample) to a Petri dish. Cut the tissue into small pieces. Weigh out a sample. One gram amounts are usually more than sufficient.

(3) Transfer the pieces to the mortar. Add 1 or 2 ml of sterile balanced salt solution to the cup. Unwrap the pestle and gently lower it into the mortar.

(4) Grasp the mortar in the left hand, the pestle in the right hand and, with the mortar resting on the bench top, grind the pestle down to the bottom of the mortar. Slowly withdraw the pestle and repeat the grinding motion until the tissue has been dispersed. Do not over-homogenize the tissue. Bits and pieces of connective tissue, etc., will not be reduced and the presence of this material should not be taken as an indication of incomplete homogenization.

   If an electrically operated unit is used to turn the pestle, you must lift the mortar upward to the suspended pestle. Heat will be generated by the shearing action so it is recommended that ice water in a metal beaker be used to support the mortar as it is raised and lowered during the shearing.

(5) Gradually add sterile buffered fluid (diluent) in proportion to the dilution wanted, i.e.,

$$\begin{array}{l} 1 \text{ g sample of tissue} \\ \underline{+\ 9 \text{ g (or 9 ml) of diluent}} \\ 1 \text{ in 10 dilution of tissue} \end{array}$$
(w/w—weight to weight or
w/v—weight to volume)

(6) Transfer the homogenate to sterile centrifuge tubes and centrifuge in the cold at 205 g (gravity) for 15 min. This may be done in non-refrigerated centrifuges by filling the cups with crushed ice and placing the tubes to be centrifuged within them. Balance the cups on a scale by adding small amounts of water to the ice-filled cups before placing them in the centrifuge.

(7) Remove the supernatant fluid with a sterile pipet. Transfer to another

centrifuge tube. Recentrifuge at 205 g for 15 min. This will contain your crude virus suspension.

(8) Remove the clarified supernate. Dispense into several small vials. Seal tightly. Label, indicating the tissue, suspending fluid, dilution, place and date of collection, and your name.

C.   USE OF MORTAR AND PESTLE

(1) Place a weighed sample of tissue into a sterile mortar. Add a sterile abrasive (sand, Alundum) plus 1 or 2 ml of a balanced salt solution.

(2) Grind the tissue and abrasive together to break down the fiber. Gradually add sterile buffered fluid in proportion to the dilution wanted. Follow the centrifugation and clarification steps as given under step 6 above.

3. Fluid Specimens

In general, cerebrospinal, blood, abdominal, pleural, or pericardial fluids require less preparation before storage than does tissue. Sterile syringes and needles are required, along with sterile receiving containers, properly labeled. Labels should have much of the same information as mentioned above.

Cleanse the area to be punctured by the needle. Gently insert the needle to the center of the vein or artery or into the cavity. Avoiding undue vacuum (pulling), gently withdraw the plunger until the desired amount of fluid has been obtained. If several samples are collected from the same animal, pool the material in a cold flask before dispensing.

All fluid samples should have sterility checks done at the time they are obtained. Refrigerated samples which are contaminated will need immediate treatment or the contamination may overwhelm the sample. Freezing of the specimens will kill most bacteria, but you should know which ones are present so appropriate antibiotics can be mixed with the materials when they are thawed for use.

Distribute the pooled material, using aseptic techniques, to several smaller containers; seal, label, and proceed with testing or store the specimens.

A.   BLOOD, SERUM, PLASMA

Whole blood may be the fluid of choice when viremia (high level of virus in the blood) is suspected. Serum will be desirable where some serologic studies are to be made or where the state of the animal before and after exposure to a virus is to be recorded. In some situations, plasma may be equally useful.

Follow the general directions given above. In all instances, it is wise to leave the needle inserted into the animal, removing only the syringe, if more blood is being obtained than one syringe will hold. A reinsertion of the needle is seldom successful.

It is best to deliver the blood from the syringe into the desired tubes or flasks with the needle removed. The compression which occurs in expelling the blood through the needle results in lysis (rupture) of the blood cells.

Venous blood specimens may be withdrawn via a Vacutainer™ system, a combination of sterile needle, syringe, and test tube assembled under vacuum. The units are available with plain, heparinized, oxalated, citrated, etc., tubes. These are easy to use; require no preparation; avoid chances of contamination, clotting, or hemolysis; are available for capacities up to 50 ml; and permit centrifuging without transfer to another tube.

For large volume sampling, the unit shown in Fig. 1-2 is easily assembled.

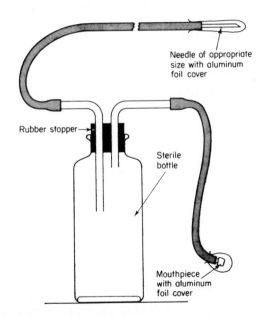

Fig. 1-2. Blood sampling unit, large volume.

(1) Wrap the stopper assembly separately; cover the flask top with foil. Autoclave both at 15 lb, 15 min.

(2) When ready to use, assemble the unit as shown. Remove the foil cover from the tubing and insert a sterile needle of appropriate size. Remove the foil cap from the short glass tubing and add a length of rubber tubing for suction.

(3) Insert the needle into the vein or heart of the animal. Apply mouth

suction as required and allow the blood to flow into the flask to the volume desired. Remove the top unit when finished and cover the flask with sterile foil.

(4) Allow the blood to clot and express the serum. If necessary, ring the blood clot by gently running a thin sterile wire around the edge. Transfer the serum to sterile centrifuge tubes and centrifuge in the cold at low speed to precipitate red blood cells.

(5) Remove the serum to sterile storage tubes, seal, label, and store.

B.   SPINAL AND ABDOMINAL FLUIDS

These receive the same general treatment as outlined above. Store fluid samples in two matching vials. The first, filled with an appropriate amount of sample, is a reserve. Place a measured amount in the second vial to dilute out for titrations or to use undiluted for study, without repeated thawing or freezing of the main reference sample. Adequate labels and sterility checks are necessary.

4. Scrapings

It is often necessary to use tissue scrapings, commonly tracheal or intestinal (i.e., the surface cells and mucosa), in virus isolation studies.

    a. Pin one end of the specimen to a dissecting pan and hold the other end of the specimen with forceps. With a sterile scalpel, scrape the surface cells loose but leave the loosened material in place.

    b. Remove the tissue with a clean, sterile spatula to a preweighed, sterile vial. Again weigh the vial and compute the weight of the tissue collected.

    c. Do a sterility check. Label and freeze the sample directly or add a measured amount of diluent and note the final dilution on the label.

5. Swabs

Samples collected this way are commonly trachea, eye, nose, or rectal swabbings.

    a. Use a sterile swab dampened (but not wetted) in a sterile diluent such as Hanks' balanced salt solution. Insert the swab into the test area and rotate it gently.

    b. Immediately deposit it into a sterile tube containing diluent and snap off the extra length of swab stick before replacing the cap.

    c. At the laboratory, remove the cap with sterile technique, grasp the swab with sterile forceps, agitate the swab in the diluent, press out the fluid against the wall of the tube, and remove the swab. Use the swab to make a sterility check.

d. Centrifuge the fluid in the cold at 1950 g for 30 min. Remove the supernate, add antibiotics, place in sterile vials, seal, label, and freeze.

6. Stools

a. Make a 20% solution of the stool sample in Hanks' balanced salt solution contained in a flask with glass beads (if the sample has been collected by swabbing, follow the directions given above). Stopper the flask and shake it vigorously for 30 min.

b. Centrifuge at 1950 g for 30 min at refrigerated temperature. Remove the supernate to another sterile centrifuge tube and recentrifuge for 30 min.

c. Remove the supernate with sterile technique, add antibiotics, place in sterile vials, seal, label, and freeze.

7. Embryonating Egg and Tissue Culture Samples

The harvesting and storage of embryonating egg-propagated or cell-cultured viruses is discussed under these headings elsewhere in this text.

C. STERILITY CHECKS

A variety of standard bacteriologic media should be available to determine if bacterial contamination has occurred (Chapter 9, Section J.). Generally nutrient agar slants and thioglycollate broth should be used. A cloudiness in the broth at the surface indicates the presence of aerobes and a cloudy layer beneath the surface with a clear layer at the top indicates the presence of anaerobes. The agar slant provides some chance to note colony characteristics of aerobes. If it is necessary to determine the identity of the contaminant, a competent bacteriologist should be consulted. For the virologist, however, the fact that a sample has bacterial contamination is usually sufficient to have him decide between using antibiotics or discarding the sample. After antibiotic treatment, however, a second sterility check is necessary and if this proves to be positive (bacterial growth is evident), it is strongly recommended that the sample be discarded.

D. SUBMITTING TISSUE FOR HISTOLOGIC EXAMINATION

Unless you are going to process the tissues yourself, you should discuss the fixatives and stains which will be used on your sample with the pathologist responsible for the microscopic examination. Since the specimen to be submitted for such examination is usually a part of that which was collected for virologic study, you must be prepared to have the proper fixative on hand at the time the animal is necropsied. The general rule is that there should be 20 volumes of fixative for each volume of tissue taken. The preparation of several fixatives is given in Chapter 9. The piece of tissue taken for histologic examination seldom need to be larger than a cubic centimeter on any edge. This piece should be taken from that area of the tissue which has not been damaged by forceps pressure or unintended cuts with the necropsy instruments.

E.   STORAGE OF SPECIMENS

Obviously it is best to test immediately the materials collected, avoiding storage except for a portion set aside for a reference specimen. Where storage must occur, the precautions discussed next should be considered.

Generally it is better to store some of the infected samples as such than to release the viruses by processing the material (grinding tissues, making suspensions, diluting collected fluids) and then storing the treated materials. In the suspended state, viruses are thought to be more subject to the detrimental conditions of pH, media, and osmotic changes than when left within the protective environment of the sample in which they have multiplied. Some of the material should be stored "as is", while another portion is processed for study. The choice must then be made as to the proper method of preserving this reference sample. There are three common methods of preservation: (1) deep-freezing, (2) lyophilization (freeze-drying), (3) storage in sterile 60% glycerol.

1.   Short Term Storage (up to 24 hr)

Where samples contain a virus whose identity is not yet determined, it may be best to hold the material at temperatures close to, but not at, freezing. The tests should begin promptly or, at the most, with a 24-hr delay but no longer. Where extreme precautions against loss of virus is necessary, the original material may be divided: some frozen or preserved by the methods described under long-term storage, and some held at cold but not freezing temperatures. This splitting of the materials into two parts makes possible the testing of several replicates of the sample. This may be important where one is dealing with viruses which are classified as highly cell associated. Such viruses have been reported to be lost when the tissues containing them were stored, regardless of the temperatures or methods used. Some attribute this to viruses not yet fully matured and which require intact cell membranes to complete their growth. Certain processes release these immature viruses as well as destroy the membranes of the cells in which they were growing.

The viruses, therefore, cannot complete their maturation cycle without these intact cells.

For that portion of the sample to be held at low temperatures, the containers used, usually glass or plastic, should be resistant to temperature stress and designed to permit an airtight seal. Cotton stoppers are not suitable! Place the tissue pieces or fluid samples in the containers; label, indicating the identity of the virus, its source, the date, and your name; and store some tubes at refrigerator temperature (4°C) while other replicates of the sample are stored under other conditions as desired.

2. Long Term Storage

A. FREEZING

Temperatures of −20 to −100°C are considered essential in maintaining viruses for periods beyond 24 hr. Containers should meet the requirements mentioned under Section 1 above. As under short-term storage, some of the tissue should be frozen unprocessed. Tissue should be clipped into pieces (0.5 cm or less) small enough to be placed in the containers. Fluids can be placed directly in the containers. Several replicates of the material should be stored. Do not freeze only one tube of the material or put all of it in a single container. Damage during storage and handling could result in the loss of the only reference source held in reserve. Although most bacteria will not withstand freezing temperatures, it is best to know the type and degree of contamination present. For this reason, bacteriologic checks should be done before the samples are frozen.

The suspending medium selected for that portion of the sample to be processed before freezing should have a certain minimum level of protein in its composition. At least 1 %, and preferably amounts up to 50 %, has been found useful in stabilizing viruses during low temperature storage. The protein may be in the form of serum, tissue particles, or egg or bovine albumin.

The temperatures which can be attained by various equipment can range from −20 to −257°C. Electric freezers have a range of −20 to −100°C, dry ice chambers are usually at −76°C, and liquid nitrogen holding tanks can attain temperatures as low as −275°C. As mentioned previously, all ampuls and tubes used to store specimens at such low temperatures must be temperature shock resistant. Where dry ice is used, the containers must also be tightly sealed since dry ice converts to CO_2 and rapidly lowers the pH of materials exposed to it. When removing such containers from low temperature chambers, it is wise to place them in a refrigerator or other cool place until they begin to attain equilibrium with the higher temperatures of the laboratory. Rapid transfer from a freezer to room temperatures (and especially the temperatures of the hands) can cause fracture of the container.

B.   LYOPHILIZATION (FREEZE-DRYING)

Place specimens in vials or tubes with necks narrow enough to accommodate the tubing on the freeze-drying apparatus. The samples used may be either virus suspensions or tissue cut into very fine pieces. In the latter instance, the freeze-drying time will be longer than when suspensions are used since water trapped in the tissue is sometimes more difficult to remove.

The sample should not occupy more than 10% of the tube's capacity. Record the weight or concentration of the original specimen. Quick-freezing of the material must precede the freeze-drying phase and this is commonly done by swirling the vial and its contents in a supercooled bath of ethylene glycol, absolute alcohol, or other liquids to which dry ice has been added. This will bring the temperatures to below −30°C. After the tissue has been frozen, transfer the vial to the freeze-drying outlets, connect them as described by the manufacturer, and then turn on the vacuum system. Sublimation of the ice leaves behind the dried residue containing the solids of the tissue and the contained virus.

Certain precautions should be kept in mind when using freeze-drying as a method of storage. Since it is only the liquid phase of the sample which is being removed, all salts and other constituents of the suspending medium are left behind. These materials, in what will then be very high concentrations as compared to the original, can be very detrimental to the virus. Failure to retain viruses using this system may be more often attributed to problems of the suspending medium than is commonly thought.

C.   60% GLYCEROL

A 60% solution of glycerin in water can be mixed with equal parts of virus suspensions or used to store tissue pieces. Combine tissue pieces with glycerin in a 1:5 ratio (weight to volume). The pieces need not be smaller than 0.5 cm cubes. After being cut, place the tissue in the glycerol solution, seal the vial, and store at the coldest temperature available. This method is usually used when access to storage temperatures of less than −45°C is not possible. This is one of the older methods of preserving tissues and has proved very useful where other storage facilities did not exist, especially the transportation of certain viruses from remote field outbreaks to laboratories many miles away. It cannot be presumed that all viruses, however, will survive equally well using this method.

D.   REFRIGERATION

Although refrigeration is usually associated with short-term, temporary storage, it should be remembered that bacteriophage (bacterial viruses) are maintained in this way. It is not uncommon to have reference cultures of phage maintained in good condition for as long as a year. Those who do maintain such cultures usually make a transfer to a fresh culture of bacteria about once a year to assure the continuance of the strain.

E. ROOM TEMPERATURE

It is obvious that many viruses survive exposure to temperatures of a wide range. Human infections, as well as those among other living forms including plants, are affected by viruses which maintain viability in excreta, cell debris, dead animals, etc. The most outstanding example of this ability to survive is found in Jenner's preservation of vaccinia virus by the simple method of retaining the scabs from infected patients and storing them in jars after air-drying the specimens.

3. Precautions

Two additional problems associated with freezing or freeze-drying are found in the thawing and the addition of reconstituting fluids. It appears that concentrated viral suspensions, cooled slowly or rapidly and thawed slowly or rapidly, lose less activity than do highly dilute suspensions [21, 41].

Monovalent and divalent cations stabilize certain viruses. RNA-viruses are protected from thermal inactivation by magnesium ions; DNA-viruses become more thermolabile. Physiological saline plus stabilizer, balanced salt solutions, and water plus $MgCl_2$ appear to be preferred for rehydration of viruses. Water would not be the fluid of choice.

It must be obvious that changes in viral activity can occur at any stage: during storage, during thawing, or during rehydration. No one method appears adequate for all viruses. In general, it appears that viruses are more stable in the less purified state; pure viruses (viruses isolated from their tissue and in a suspension from which cell materials have been removed, usually by centrifugation) are more stable in concentrated suspensions when frozen at temperatures below $-170°C$ and when reconstituted in a balanced salt solution.

F. HANDLING PURIFIED VIRUSES

In the procedures discussed above, viruses, for the most part, were either in tissue or suspended in fluids (spinal, blood, tissue homogenates) whose composition is unknown. Partially purified or purified virus preparations present additional problems of preservation since the protective menstruum has been removed. Then the method of storage becomes even more critical.

TABLE 1-1

The Effect of Temperature and Media on the Storage of Some Viruses

Nucleic acid type	Group designation	Virus strain	Treatment			
			Non-frozen, stored 30 days at 4°C	*Frozen at −40°C, stored 30 days at −65°C*	*Frozen at −76°C, stored 30 days at −65°C*	*Lyophilized at 0°C, stored 30 days at −65°C*
				Titration decreases		
RNA	Myxovirus	Respiratory syncytial	Extreme BSM > DMSO > PDA	Extreme	Extreme	BSM—marked PDA—no change
		Influenza	"	No change	No change	"
		Measles	"	Slight decrease BSM > DMSO PDA—no change	Moderate	BSM—marked PDA—moderate
		Rubella	Marked BSM > DMSO > PDA	BSM, PDA—marked DMSO—moderate	Moderate	"
	Picornavirus	Polio	No change	No change	No change	Marked BSM > PDA
		ECHO 9	No change	No change	No change	Marked PDA > BSM
DNA	Herpes	Herpes simplex	Slight	No change	No change	No change
	Pox	Vaccinia	Slight	No change	No change	BSM—slight PDA—no change
	Papova	SV-40	Slight to moderate DMSO > PDA > BSM	No change	No change	BSM—moderate PDA—slight

BSM: Medium 199; physiological saline.
DMSO: Medium 199 + DMSO to 2% final concentration.
PDA: Medium 199 or physiological saline + PDA (calcium lactobionate + human serum albumin); 1% of each.

Not only the suspending medium but the salt concentration, temperature of freezing, storage temperature, and rates of freezing and thawing become paramount.

It has been found that there is a variation in the ability of some viruses to withstand storage conditions which are well tolerated by other viruses. In the work of Rightsel and Greiff [41] not only were some viruses found to be temperature sensitive, but the suspending media used at certain low temperatures was found to have a marked effect on the viability of the viruses under study (Table 1-1). Such information should alert those anticipating the storage of valuable reference strains of virus. No one medium or system will be applicable to all circumstances, and several may have to be tried before the most suitable is found.

Although the problems cited above do exist, most changes and chemical activities are negligible at temperatures of −130°C or below. Sera, enzymes, and many viruses can be stored in ultralow temperatures without damage: in −70°C freezers, in the gas phase over liquid nitrogen (−130°C), or in liquid nitrogen at −257°C.

The problems which appear when viruses are lyophilized are similar to those encountered when the materials are stored at freezer temperatures. Attention to the medium used at each temperature level seems to be of great importance, and members of the same virus group cannot be expected to respond alike.

Egg Propagation of Viruses

TWO

Embryonating chicken eggs were first used in the propagation of viruses by Rous and Murphy, 1911 [43] with studies on a transmissible tumor. An interesting review on the very early use of eggs for virus propagation was published by Goodpasture in 1938 [19]. Since that time and until the mid-1950's, almost all research and testing with viruses depended on animals or embryonating eggs for the source of living tissue.

If eggs are compared to the two other sources of living tissue (experimental animals and cell cultures) usually used in viral propagation, the following advantages become obvious: they do not require the care, facilities, or space of experimental animals nor the technical skill and time necessary for the preparation and maintenance of cell cultures.

You must recognize, however, that the cost per egg (which must include the cost of producing the egg) can be more than that for cell cultures made in the laboratory. The embryonating egg also takes more incubator space than the same number of cell culture tubes. Like experimental animals, the infected specimens must be kept isolated from those yet to be used in propagation and testing.

The embryonating egg is, nevertheless, a convenient and easily manipulated source of living tissue for the propagation and testing of viruses. This section will cover the proper method of selecting and preparing eggs and methods for inoculating, harvesting, and storing virus-infected tissues obtained from inoculated eggs.

A. FACILITIES AND EQUIPMENT

1. Incubators

Small commercial incubators can be purchased which are equipped to supply the balance of humidity, air circulation, and heat. Where the purchase of such equipment is not feasible, conversion of a small room or use of a standard incubator is possible. Generally, incubation of embryonating eggs should occur under the following conditions: temperature, 37.5°C; humidity, 62%; forced air (fan) circulation to ensure a balanced temperature; and automatic turning racks to rotate the eggs during incubation.

2. Candling Area and Equipment

A room which can be darkened is essential. This room can be relatively small, for it need only accommodate the person doing the candling, have about 4 to 5 ft of bench space on which to work, and provide adequate ventilation. If this is not possible, construct a divider setting off part of a laboratory bench in a dimly lit corner or install a black photographer's curtain at ceiling height suspended from a wall-mounted rod.

Equipment needs are primarily a suitable light source and apparatus to make small holes in the shell of the egg. Candling lamps are available commercially (Appendix B) and provide the best focused light source. Substitutes, however, can be constructed (Chapter 9) or a detachable dissecting microscope lamp with a variable rheostat is an excellent choice. Either an egg punch or drill, commercially available, can be used in making the shell holes (Chapter 9, Section K). Previous to the development of the egg punch, hand-held dental drills were standard equipment in most laboratories. These needed constant care in the changing of oil and drill tips and became quite warm if prolonged drilling was necessary. Since the egg punch is nothing more than a solid metal bar with an appropriately formed punching tip and since it has no moving parts, it is both economical and an efficient tool for making the shell holes.

3. Inoculation Area and Equipment

The ideal inoculation area, like that for candling eggs, is a small room which need not exceed 5 ft in length. It should have lab bench space and an area for an equipment cart. Ultraviolet lighting with reflectors to protect the worker's eyes is a very efficient method of attaining a clean room atmosphere.

Where facilities permit, several inoculation areas should be constructed; it is preferable that each type of virus be handled in isolation from others. This avoids the danger of contaminating one strain of virus with another.

Under no circumstances should the inoculation or harvesting of one virus occur in a room where others have been handled recently unless scrupulous cleaning of the surfaces of the entire room (walls, floor, workbench, etc.) has been done. Such facilities are generally not available or necessary in a class situation where only an open room and a single virus sample provides the first experience in virology.

B. THE EGG SUPPLY

Certain information about the supply flock should be part of the experimental data. Results of tests for common viral diseases such as infectious bronchitis (IBV), Newcastle disease (NDV), pleuropneumonia-like organism (PPLO) infections, and others should be known. These conditions affect the quality of the eggs and introduce latent agents, bacteria, and viruses which may be carried by the developing embryo. A complete discussion of such tests is beyond the scope of this text but should be part of the information of anyone seriously attempting research with the embryonating egg as the source of living tissue. For this reason it is best to purchase eggs from a single producer who can then supply the necessary information. The following points should be part of the experimental record kept for all eggs used.

1. Flock Immunization History

    a. Type of vaccines used, manufacturer, serial number, and date administered.
    b. Test results on evaluation of immunization.
       (1) Test method used for evaluation.
       (2) Results, including number of birds tested.

2. Production Records

    a. Egg production over history of laying period.
    b. Feed conversion records.
    c. Hatchability records.

C. ANATOMY OF THE EMBRYONATING EGG

For those who have little, if any, background in embryology, a brief outline of common terminology is provided. Diagrammatic figures assist in understanding the anatomy of the embryonating egg as it is used in the propagation and testing of viruses (Fig. 2-1).

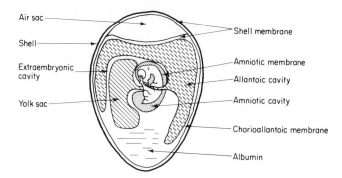

Fig. 2-1. Embryonating egg.

1. The Embryonating Egg

This term identifies any egg within which an embryo is developing. It is common to refer to fertile eggs which have been incubated as being 3, 5, 10, etc., days old. This does not refer to the time that has elapsed since the egg was laid but to the time that is has been incubated. Although certain, early, primitive developmental stages occurred after fertilization and while the shell was being formed around the egg contents, this development does not continue after the bird has laid the egg. Unless the egg is subsequently incubated, either naturally or artificially, further development of the embryo is suspended. This potential is not lost so long as the eggs are not subjected to extremes of temperature. It is this characteristic of suspended development that makes it possible for the avian species to produce, in their natural habitat, a *clutch* of eggs before settling down to incubate them. It is this same principle of controlled storage which makes it possible for commercial hatcheries to hatch thousands of chicks on a specific day.

2. Shell and Shell Membrane

If, using sharp-pointed forceps, you gently pick away the shell of an egg, a thin white membrane will be found. This membrane is so closely attached to the shell itself that if broken pieces of shell are examined, most would find it difficult to accept the fact that there is a membrane lying tightly up against the inside of the shell. Scraping of the inner surface of the shell with forceps will show this to be true. This is the shell membrane.

The shell itself is only one of several systems of the avian egg which functions as an exchange system. Across its surface gaseous and liquid molecules pass in both directions. This is why the incubation of the egg must occur in the presence of humidity as well as adequate air circulation. If an egg is incubated in an atmosphere too low in humidity, the egg will lose moisture, the contents will become dehydrated, and death of the embryo will result. The same is true of air circulation. If embryonating eggs are tightly packed for too long a period of time while under incubation, impairment of air circulation will also result in embryo deaths. The shell and shell membrane of an egg, therefore, are not just inert coverings of a living content, but themselves function to maintain the life form within.

3. Air Cell

All normally formed eggs have a rounded end and opposite this is the end formed by more acute angles of the shell, i.e., the *pointed* end. Under the shell and shell membrane of the blunt, rounded end is an empty space. The egg contents are prevented from entering this area by a sturdy layering of several membranes which act as a barrier. The space, called the *air cell*, functions in both respiration and pressure adjustments.

4. Chorioallantoic Membrane (CAM) and Chorioallantoic Sac (CAS)

Attached to the developing embryo and originating from the hind gut is a sac-like structure which functions to remove soluble, insoluble, and gaseous waste products. As the embryo increases in size, the sac also enlarges until it surrounds the embryo much like a double-layered umbrella. The chorioallantoic membrane which forms the sac is usually called the CAM. The membrane, together with the contents of the sac, are often designated simply as the CAS (chorioallantoic sac). The fluid contained in the sac is known as *chorioallantoic fluid* (CAF) or, more simply, AlF. It is necessary to differentiate between a CAM inoculation, where the intent is to place viruses on or in the membrane, and CAS or allantoic fluid (AlF) inoculations, where the intent is to place the virus inoculum in the fluid within the space formed by the membrane, i.e., the CAS.

5. Yolk and Yolk Sac (YS)

Not only is there a chorioallantoic sac extending from the developing embryo, but a similar structure called the *yolk sac* (YS) completely surrounds the yolk contents of the egg. Through its membrane and blood vessel system, the nutrients of the yolk are transferred to the embryo. As the embryo matures, the contents of the yolk sac noticeably diminish until the sac is 1 cm in diameter approximately 3 days before the embryo hatches. At this stage, the embryo has attained such size that it nearly fills the egg and, except for a small opening in its abdomen from which the yolk protrudes, the embryo is perfectly formed. In the final days remaining to hatching time, the opening in the abdomen will slowly close around what remains of the yolk sac which, with the much reduced contents, will be absorbed into the digestive system of the embryo.

6. Amniotic Sac (AMS)

This very thin membranous system is most easily seen in embryos of 4 to 9 days of age. It appears as a translucent, clear, liquid-filled dome stretched over the developing embryo. In fact, it completely surrounds the embryo except for the areas where the CAS and YS are attached. Its involvement with the embryo and its function can best be seen by gently removing the entire contents of the egg into a Petri dish and examining the intact structure. This membrane and its contents serve to protect the embryo against physical damage and function as an area of exchange. As the embryo enlarges, the membrane stretches to accommodate the increase in size and serves less and less in the role of a buffer against physical motion. By the time the embryo has matured, the membrane is thinly stretched and barely visible as it surrounds the fully developed form.

D. SELECTING EGGS FOR VIRUS PROPAGATION

1. Determination of Viability

One of the purposes of candling eggs is to determine the viability of the embryo. Generally, a healthy embryo will impart an orange-yellow color to its egg because of the blood vessels and circulating blood. An embryo which is dead or dying will be easily detected because of diminished blood vessels or their complete absence. If the embryo has grown large enough nearly to fill the egg, lack of motion may be your only clue to death. An egg containing a small dead embryo will usually have a clear yellow color. Any coloration tending to green or black is an indication of extreme contamination, and such eggs should be carefully removed and placed in a discard container. In addition, the contents may have the consistency of water rather than the viscous character of a healthy egg.

2. Age of Embryonating Eggs and Suitability to Various Inoculation Routes

The age (incubation time) of embryos best suited to the propagation of various viruses and the routes of inoculation commonly employed are given in Table 2-1. There are, generally, only limited times when an embryonating

TABLE 2-1
Propagation of Various Viruses in Embryonating Eggs

Virus	Age of embryo (days)	Advisable route of inoculation	Temp (°C)	Incubation time	Reaction	Suggested harvest material
Infectious bronchitis virus (IBV)	5–6	YS	37	3–4 days	Deaths and/or abnormalities	Yolk, YS
	9–11	CAS	37	3–7 days	Deaths and/or abnormalities	AIF, CAS
Newcastle disease virus (NDV)	9–11	CAS	32	4 days	Deaths, HA*	AIF, CAS
Fowl pox	10–13	CAM	37	3–5 days	Pocks	CAM
Laryngotracheitis (ILT)	9	CAS	37	3 days	Deaths	AIF, CAM, embryo
	10–13	CAM	37	3 days	Pocks	CAM
Herpes simplex	10–12	CAM	37	2–6 days	Pocks	CAM
Vaccinia	10–12	CAM	37	2–3 days	Deaths, pocks	CAM
Variola	10–12	CAM	37	3 days	Pocks	CAM
Rous sarcoma virus (RSV)	9–11	CAM	37	5–10 days	Pocks	CAM
Eastern equine encephalomyelitis (EEE)	10–12	AM, CAS	37	18 hr	Deaths	AIF, AmF, CAM, embryo

*HA = hemagglutination; i.e., the ability to agglutinate various red blood cells.

egg can be used for the propagation of certain viruses because of virus selectivity for certain membranes or certain stages of development of the embryo. Little success will follow if attempts are made to substitute routes of inoculation or incubation ages of the embryo.

3. Determination of Embryo Position

The embryo's position must be determined so inoculation routes into specific membranes or areas can be accurately placed. Choose positions which will place the inoculation route a little to one side or the other of large vessels, still keeping as close as possible to the desired route.

4. Efficiency in Candling

It is generally more efficient to make the appropriate hole in the shell of the egg at the time it is candled. The exception to this may be when an unusually large number of eggs are found to be dead. Then it may be best to candle all the eggs first, remove the dead or dying ones, recandle, and then make the desired holes.

5. Procedure

Hold a candling lamp securely against the blunt end of the egg. The light should illuminate the air cell and the egg contents. After a hole has been "punched" or drilled in the shell, each egg should be numbered on its side with a pencil. In this way, the inoculum for each egg can be recorded, and deaths or other observations pertinent to that particular embryo can be noted. Figure 2-2 is sample pages from an egg inoculation record book. Observe the manner of recording the information required for the various tests. All eggs should remain in a vertical (blunt end up) position, except those prepared for CAM inoculation. Place these eggs in a horizontal position in the egg flat. This will require extra flats to accommodate the same number of eggs.

A.   FOR YS ROUTE

Eggs of 5 to 6 days of age.
(1) Rotate the egg in its tray until blood vessels can be seen close to the margin of the air cell and within the yolk-filled area of the egg. These vessels may appear as nothing more than an array of faint lines, orange in color, extending from a clear halo. The developing embryo, surrounded by its fluid-filled sac, is within the area of the halo [Fig. 2-3(a)].

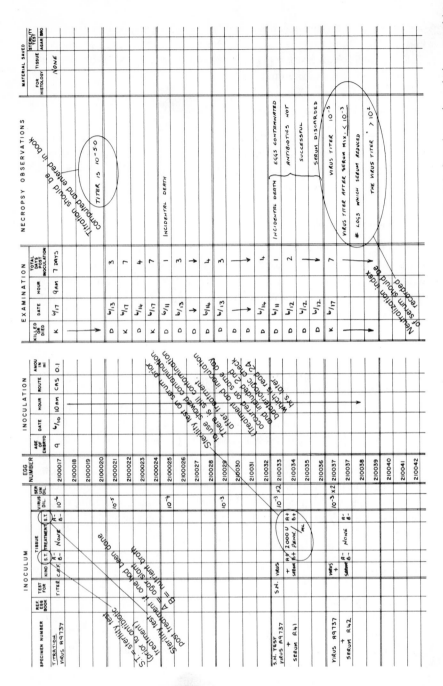

Fig. 2-2. Sample egg inoculation record book.

24

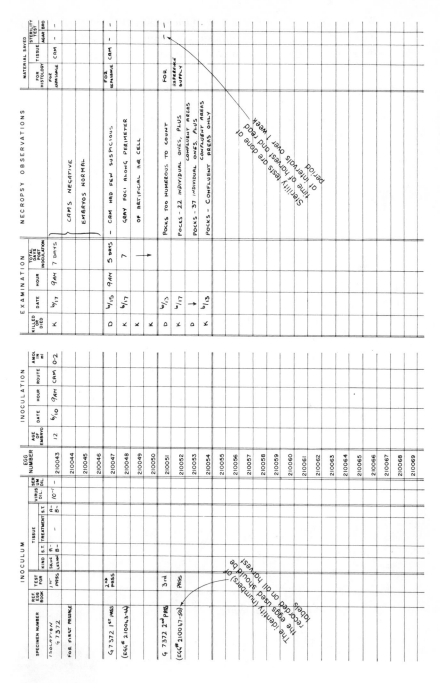

Fig. 2-2. Sample egg inoculation record book (cont.)

(2) With an egg punch, make a hole in the shell (but not through the shell membrane) in a position opposite the embryo and halfway between the margin of the shell membrane and the top of the egg [Fig. 2-3(b)].

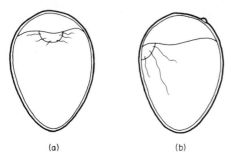

(a)   (b)   **Fig. 2-3.** Yolk sac route.

B.   FOR CAS ROUTE

Eggs of 9 to 11 days of age.
(1) Rotate the egg, blunt end up, in its tray until blood vessels, combined with a darkened shadow, indicate the position of the embryo [Fig. 2-4(a)].
(2) Continue to rotate the egg away from you (either direction) until the embryo is a full one-third turn from facing you directly [Fig. 2-4(b)].
(3) Punch a hole in the shell $\frac{1}{4}$ in. above the margin of the air cell membrane and one-third the distance around and away from the embryo [Fig. 2-4(c)].

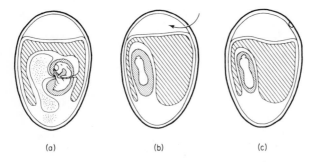

(a)   (b)   (c)

Fig. 2-4. CAS route.

C.   FOR CAM ROUTE

Eggs of 10 to 12 days of age.
(1) In addition to the candling lamp and egg punch, assemble the following:

A 1- or 2-ml sterile pipet
A cotton-tipped swab (not sterile)
Extra egg flats
A 2-ft length of soft rubber tubing, approximately $\frac{1}{2}$-in. diameter
Two sterile syringes (capacity not critical, used to hold needle
  only)
A 26- or 28-gauge needle, $\frac{1}{2}$-in. length, sterile (gauge and
  length not critical; used to probe shell membrane for admis-
  sion of saline)
A 26- or 28-gauge needle, non-sterile, to puncture the end hole
  in preparing the artificial air cell
Tube of airplane (hobby) cement
Weak tincture of iodine
Sterile balanced salt solution

(2) Candle the eggs and punch a hole directly in the top of the air cell
[Fig. 2-5(a), (b)]. Punch a second hole in the side of the egg directly
over the area of the greatest blood vessel development. This is the
area occupied by the embryo [Fig. 2-5(c)].

(3) With the sterile needle (need not be mounted on a syringe), puncture a
hole through the shell membrane at the blunt (air cell) end of the egg.

(4) Lay each egg horizontally in the tray, with the hole in the side of the
egg facing up [Fig. 2-5(c)]. Turn on room lights.

(5) Decontaminate the area around the hole in the sides of all eggs by
swabbing with tincture of iodine. Make the swabbed area about 2
cm in diameter around the punched holes [Fig. 2-5(d)].

(6) Put two to three drops of saline on all the "side" holes so far prepared [Fig. 2-5(e)]. Gently abrase the shell membrane with a needle
attached to an empty syringe. Let the saline seep onto the chorioallantoic membrane beneath. This helps loosen the CAM from the
shell membrane much as would a lubricant.

(7) Turn room lights off and candling lamp on.

(8) Hold each egg horizontally against the candling lamp, swabbed hole
still facing up. Place one end of the rubber tubing in your mouth
and the other end securely against the hole in the blunt end of the
egg [Fig. 2-5(f)].

(9) With gentle suction, draw the contents of the egg into the area previously occupied by the air cell. This will cause an air cell to develop
beneath the hole punched in the side of the shell as the egg contents
shift into the air cell. In doing so, the saline and room air are drawn
into the egg [Fig. 2-5(g)].

Occasionally, embryos will revert to their natural position; i.e.,
they will shift from the false air cell back into the usual position.
This may happen soon after preparation or sometime during incubation after inoculation. The only measures which can be taken to
discourage this is good sealing of the two holes at the appropriate

times, i.e., the hole in the blunt end of the egg immediately after preparation, and the hole in the side of the egg as soon after inoculation as is practicable.

(10) Place the eggs in the flat in a horizontal, side hole up position. Seal the hole in the blunt end of the egg with a drop of glue.

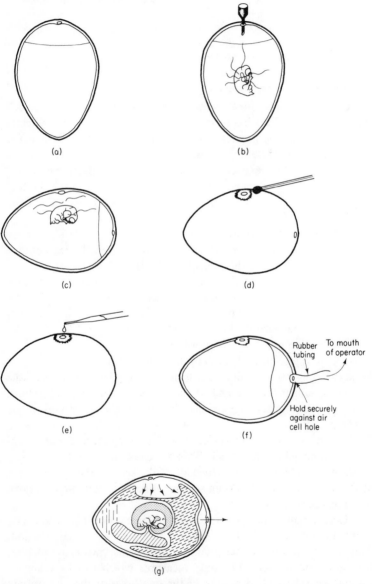

(a)

(b)

(c)

(d)

(e)

(f)   Rubber tubing   To mouth of operator

Hold securely against air cell hole

(g)

Fig. 2-5. CAM route.

6. Preinoculation Storage of Eggs

You may prepare the holes for CAS and YS inoculations and then store the eggs for several hours, or overnight, before inoculation. This is not the preferred situation but becomes a common procedure where large numbers of eggs are to be prepared. If shell membranes have not been punctured in the preparation of the hole, they can serve as a barrier to common contaminations. Humidity plays an important part in the ability of this membrane to exclude contaminants. If, after the holes are punched, the eggs are stored in an incubator which has low humidity, the shell membrane will dry and split and the egg may become contaminated. If it remains intact and the area is swabbed with tincture of iodine just prior to inoculation, little, if any, contamination should result. Because of the exposed membrane on the CAM-prepared eggs, such eggs should be inoculated within hours of preparation. Store eggs prepared for inoculation separately from those which have been inoculated. Again, the ideal cannot always be attained, and little difficulty has been reported where inoculated and uninoculated eggs are placed in the same incubator.

7. Inoculation of Eggs

One common caution for all routes of inoculation is to avoid too large an inoculum. It is seldom advisable to use more than 0.2 ml. Variances may occur with some routes; i.e., you can put more into a yolk sac (a large area of the developing embryo) but certainly not into the amniotic sac since internal space is limited and rupture of delicate tissues may occur.

A. YS ROUTE

Use a 22-gauge, $1\frac{1}{2}$-in. length needle, combined with a syringe of sufficient capacity to hold the prepared inoculum. The length of the needle is suggested only because the purpose of the yolk sac route is to place the inoculum below the embryo and within the yolk material for rapid absorption. If one visualizes the point where the inoculum is to be deposited, there is no serious disadvantage to the use of longer needles. You simply insert the needle straight down into the yolk sac until its point is one-third to one-half the depth of the egg (Fig. 2-6). A needle which is too short, however, will not permit the deposition of the inoculum within the yolk sac. The gauge of the needles is dictated by the length and, therefore, the longer length needles are usually those with the larger bore (gauge). Such a needle has use also in harvesting material from the yolk. This material is viscous, and a narrow gauge needle would not allow the yolk material to pass through. Thus, the purchase of this gauge needle in the length mentioned is a matter of economic efficiency since the one type will serve both purposes.

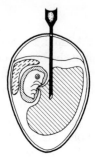

Fig. 2-6. Yolk sac inoculation.

B. CAS ROUTE

Since the introduction of the inoculum into the egg for this route does not require deep penetration of the needle, a shorter one is generally used. The 26- or 28-gauge needle is preferred because of the fineness of the point. Because the area to be inoculated is within the thin membrane of the CAS, the fine gauge needle permits inoculation with the minimum amount of trauma. It is inserted into the shell hole at a 45° angle from the perpendicular and is inserted into the egg about $\frac{1}{2}$ in. total depth (Fig. 2-7).

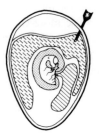

Fig. 2-7. CAS inoculation.

C. CAM route

A thin bore needle minimizes the trauma of inoculation. Slip the tip of the needle under the shell membrane which was already split in admitting the saline used during the preparation and gently deliver the inoculum onto the surface of the chorioallantoic membrane. If the inoculum flows back out of the hole, it may be due to (1) failure to insert the needle tip sufficiently below the surface of the shell membrane or (2) shifting of the embryo, causing reduction in the size of the artificial air cell. The solution to the first is obvious. In the second instance, you can attempt to reestablish the artificial air cell, although it is preferable to select another egg for use (Fig. 2-8).

Fig. 2-8. CAM inoculation.

E. EXAMINATION SCHEDULE

Inoculated eggs should be examined (candled) with some frequency until a determination is made concerning the length of time necessary to observe the effect of the inoculum. The preferred schedule generally functions as follows.

1. Make the first examination approximately 16 to 18, but never later than 24 hr, after inoculation. This examination is to find eggs which are dead or dying from the trauma of inoculation. Some of the eggs may die within this time period due to the virus inoculum. However, the balance of the eggs, also inoculated with virus suspension, will provide sufficient material to harvest (collect) for repassage or further study. It is not generally necessary to be concerned about saving materials from these early deaths. These eggs are always best discarded. If, for some reason, an overwhelmingly concentrated virus suspension was inoculated which killed the eggs within this "incidental death" period, the best approach is to return to another sample of the original material and inoculate eggs again, but with a dilution of 1 part in 10 (1 + 9) or more.

2. Make the second as well as subsequent examinations at 24-hr intervals, at least. Where embryos are dying or dead, remove and refrigerate them immediately. Harvest materials from these eggs or examine them as soon as possible, preferably within 4 hr of their removal from incubation. Candle the eggs several times during a working day, refrigerate all the ones requiring removal from incubation, and then harvest or examine the eggs at one time.

3. The total duration of the candling period must be determined by (a) the tissue which is to be harvested, (b) the response pattern of the embryos, and (c) the purpose of the test. If yolk is to be harvested, you must remove the embryos before their twelfth day of incubation. Beyond this time, the yolk is so thickened that it cannot easily be withdrawn. The surviving embryos should still be examined for deaths or gross pathologic change. In theory, then, you can attempt candling eggs until hatching time with examination of the about-to-hatch chicks. Beyond the fifteenth day, however, little can be seen because the size of the embryo, and its opacity, makes it impossible to illuminate the interior of the egg.

F. COMMON EMBRYO RESPONSES TO
INOCULATION

Embryos can respond to infection in a number of ways. A few responses which may be noticed following inoculation of the embryonating egg are death; gross embryo abnormalities such as stunting, dystrophic muscles, and other limb or body distortions; histopathologic response discernible only under the light microscope; and lesions of the extracellular membranes such as edema or localized lesions as seen in the formation of pox. There should be no presumptions on your part that a certain virus will cause only certain pathologic responses in the embryos. A critical eye open to all possibilities will prove most profitable, not only in seeing what is obvious but in discovering what others may not have seen.

G. HARVESTING MATERIALS FROM
INOCULATED EGGS

1. Introduction

A.   INOCULATION ROUTE VERSUS MATERIAL HARVESTED

The most important aspect of the inoculation-harvesting sequence is the recognition that the route of inoculation does not necessarily dictate the tissue to be harvested. There are times when the choice route of inoculation may be the CAS route; yet the tissues to be harvested will be purely embryo-associated (heart, muscle, etc.) or a tissue which was indirectly exposed to the inoculum (yolk, yolk sac). The instructions for harvesting, therefore, are outlined according to the various tissues which might be removed, and reference to the method of inoculation will be made only in passing.

B.   PREPARATION OF THE WORK AREA

Harvesting the desired tissue should be done with as much care in the preparation of the work area as was given in the preparation for inoculation. Generally, a draft-free, closed room supplied with air filtration system and a work area which can be disinfected meets most requirements. Contaminated equipment should be kept separate from contaminated tissues and fluids. A covered pan containing disinfectant should be used for contaminated equipment (needles, syringes, forceps). For egg materials, tissues, and fluids, a separate container is advisable. If such tissue can't be incinerated, a heavy-duty plastic bag placed in the container will facilitate disposal. If the tissues must be autoclaved, line the container with several layers of foil or paper to facilitate their removal from the pan.

2. General Procedures

A. MATERIALS

The following equipment will be needed regardless of the material to be harvested. All equipment must be sterile. Under each type of tissue to be harvested will be listed only those articles needed in addition to these.

> Sterility check tubes, in rack:
>> Nutrient agar slant for most aerobes
>> Thioglycollate broth for most anaerobes
>
> Pipets, 1.0 ml, sterile
> Sharp, curved-tip forceps
> Alcohol, 70%
> Gauze squares (nonsterile): 2 × 2 in. minimum size
> Contamination pan with lid
> Container for contaminated pipets
> Sterile flask with stopper, in ice bath, for pooling the harvested
>> material
> Individually labeled, sterile vials, to receive aliquants of
>> pooled material
> Bunsen burner (or alcohol lamp)

B. METHODS—GENERAL

(1) *Selection of Embryos to Be Harvested*   Since viruses require viable tissue for multiplication, the most potent yields are obtained from embryos dying but not yet dead. This is why candling of the eggs on a frequent schedule is recommended. With some viruses, examinations should proceed at 8- or 12-hr intervals. With others, once every 24 hr will be adequate.

For efficiency, and if several examinations of embryos are performed during a day, collect those which are dead or dying over a time period, hold under refrigeration, and harvest at one time. Where a test is to be terminated although all embryos are viable, harvest tissues and fluids directly, without refrigeration. This usually results in the release of much blood into certain of the fluids—an undesirable factor when AlF is being collected or when hemagglutinating viruses are to be harvested. If embryonic tissue or yolk is being harvested, this will not be a deterrent.

Where chilling of the embryos is desired, 2 to 4 hr in a refrigerator or $\frac{1}{2}$ hr in a -50 to $-70°C$ freezer will be sufficient. This immobilizes the embryo and reduces blood flow, permitting (a) better visibility when the eggs are opened and (b) the blood-free harvesting of certain fluids.

(2) *Labels*   Typewrite all labels and affix to the storage vials before the harvest materials are placed in them. It is nearly impossible to fix labels to a chilled glass surface on which moisture from the atmosphere is condensing. Unless labels are already placed on the glass before harvesting, you may find it impossible to do so later. Transparent tape is an excellent material to use since, in affixing your label to the vial, the tape can cover and thus protect the type. But if for any reason the contents become partially thawed, the tape will quickly slide off the vial. When thawing harvested material for further study, exercise caution to avoid dislodging the label. As an alternative, use white-surfaced adhesive tape on which typescript is visible. It does not become unglued from the vial during thawing. Since typescript on such tape has a tendency to smear, this label is often overlayed with transparent tape to protect the type.

(3) *Disinfecting Surface of the Eggs*   Harvesting of material from the eggs may be done with the eggs left in·the trays in which they were incubated. Disinfect the surface of the egg with a gauze square dipped in 70% alcohol. All the eggs to be opened may be disinfected at one time. Rub the shell gently over its entire surface if the entire contents are to be removed. If only part of the shell is to be removed, then sterilize a generous area beyond the anticipated opening. The egg to be harvested may be placed on a flameproof form, alcohol applied, and then the alcohol ignited. The non-flamed system seems more to be recommended since danger to personnel is minimized, and the flame method has not been shown to be more effective than the non-flame method.

(4) *Opening the Egg*   With alcohol-flamed sharp-pointed forceps, tap the egg shell along the line marking the area to be exposed. Tap with sufficient force to break the shell without penetrating the egg. Any pieces of shell which come loose in the process should be brushed into a small container for later disposal. Disinfectant should be in the container since the inner surface of the shell will be contaminated with virus if the egg has been successfully infected. Break the shell along the margins of the natural air cell or, in the case of a CAM harvest, along the margins of the artificial air cell.

(5) *Exposing Tissue or Fluids to be Harvested*   Pull away the membrane thus exposed with sterile forceps and, with the assistance of the various harvesting equipment (discussed in each section), remove the tissues or fluids to a sterile, labeled container held in an ice bath.

(6) *Treatment of Harvested Materials*
(a) *Pooling*   You may have such a small harvested sample that it will fit into a single small vial. This is rare; usually several milliliters are obtained. Under no circumstances should the material harvested first be put into one

container, and the materials harvested second be put into another, with the labels the same on both. You cannot presume that the two are homogeneous. Therefore, it is the general rule that a vessel be used large enough to "pool" all the AlF or all the YS being harvested. When the harvest is completed, mix this material thoroughly, and then dispense to labeled vials.

(b) *Sterility check*　Do sterility checks at the time of the harvest to identify any bacterial contamination which may have occurred during harvesting, inoculation, or incubation. This is most conveniently done by having the appropriate bacteriologic media available and labeled prior to the harvest. The sample may be obtained from the tip of the pipet or forceps used in harvesting. Use a 1-ml sterile pipet to withdraw approximately 0.2 ml of fluid from the "pool" flask. Using sterile techniques, spread a small drop of this over the surface of an agar slant for aerobic bacteria. Drop a small amount into thioglycollate broth and then lower the pipet to the bottom of the tube to distribute the inoculum below the surface of the medium for anaerobic bacteria. Incubate at 37°C and examine for bacterial growth after 24 and 48 hr. If bacterial growth is observed, two alternatives are possible: (1) treat with any one of a number of suitable antibiotics or (2) discard the specimen. In the former instance, thaw the frozen specimen, add antibiotics, and incubate at refrigerator temperatures for several hours depending on the virus involved. Make another sterility check on an aliquot and refreeze the treated specimen. If the second sterility check is positive, it is best to discard the sample.

(c) *Storage*　After the sterility check has been completed, the harvested material is dispensed in previously labeled small vials and stored at −50 to −70°C.

C. METHODS—SPECIFIC

(1) *Amniotic Fluid Harvest*　The best yield is from embryos up to 15 days of maturity, i.e., up to 6 days postinoculation.

(a) *Materials*　Basic list plus sterile 2- or 5-ml syringe(s) fitted with 26-gauge needle(s).

(b) *Method*　Disinfect and remove the shell over the air space. Gently insert the syringe needle within the amniotic sac which surrounds the embryo. Withdraw the fluid very slowly to avoid attracting the delicate membrane or the embryo to the needle which will occlude the opening. Pool fluid from all the embryos. Take sterility checks, dispense, and store as described.

(2) *Allantoic Fluid Harvest*　The best yield is from embryos up to 14 days of maturity, i.e., up to 5 days postinoculation.

(a) *Materials*　Basic list plus 10-ml sterile pipets.

(b) *Method*

(i) Remove the shell over the air space of all eggs from which allantoic fluid will be removed. Puncture the chorioallantoic membrane immediately under the shell with sharp, pointed, curved-tip forceps. Make the puncture away from the embryo.

(ii) Turn the forceps so their curve is downward and the tips upward. This creates a pocket into which the fluids will flow and holds tissues away from the pipet.

(iii) Place the tip of a 10-ml pipet in the pocket maintained by the upturned forceps, and withdraw the fluids. (The amount will vary depending on the age of the embryo. The maximum yield is about 8 to 10 ml from a single embryo of 13 days maturity. Lesser amounts of AlF will be obtained from embryos younger or older.) Place the fluid in the "pool" vessel which is in an ice bath.

(iv) Repeat the process using the same forceps and pipet as long as the material being harvested is the same and sterile techniques are followed. Pool all similar material. Take sterility checks, dispense, and store material as described.

(3) *Amnioallantoic Fluid*   The maximum yields are from embryos up to 14 days maturity; i.e., up to 5 days postinoculation.

(a) *Materials*   Basic list plus 10-ml sterile pipets.

(b) *Method*   The procedure for amnioallantoic fluid is the same as for allantoic fluid except that the amniotic membrane surrounding the embryo is also ruptured to release the amniotic fluid into the surrounding area.

(4) *Yolk*   The maximum yield is obtained from embryos up to 11 days maturity, i.e., up to 6 days postinoculation.

(a) *Materials*   Basic list plus sterile 5-cc syringe and 16-gauge $1\frac{1}{2}$- to 2-in. needle.

(b) *Method*

(i) Withdraw the yolk by simply inserting the needle straight down through the inoculating hole directly into the yolk sac and into the center of the yolk contents. Withdraw the plunger gently to minimize the pulling of the sac into the syringe (this will obstruct the needle).

(ii) Pool all material desired. Do sterility checks and dispense and store materials.

(5) *Yolk Sac*   This tissue has developed sufficiently to be harvested by the tenth day of incubation. Essentially it can be harvested through the fifteenth day but after that time the embryo's absorption of the yolk and the yolk sac makes harvesting difficult.

(a) *Materials*

      A sterile tissue homogenizer; either mortar and pestle or Ten
        Broeck tissue grinder
      Hanks' medium, sterile
      Scissors
      A sterile beaker or flask
      A scale

(b) *Method*

(i) Open the egg, and expose the yolk sac. Grasp the sac with forceps and, while lifting, clip the sac away from the surrounding tissue. You may dump the entire egg contents into a Petri dish and then remove the yolk sac but this increases the glassware used and encourages contamination.

(ii) Place the piece, or pieces thus obtained, in a sterile wide-mouthed beaker or flask until a sufficient amount has been obtained. Weigh the tissue and container.

(iii) Transfer the tissue to the homogenizer with sterile technique. Note the difference in weight of the container after the removal of the tissue and calculate the weight of the harvest you will homogenize.

(iv) Homogenize the tissue (Chapter 1, Section B).

(v) Add sufficient medium to make a stock suspension of the harvested material, making note of the ratio of tissue to medium on the labels of all tubes, i.e., YS 1 + 4 in Hanks' medium. Mix the homogenized and diluted harvest material well. Do sterility check. Dispense into sterile vials. Freeze.

(6) *CAM*   Harvest from embryos up to 15 days of maturity; i.e., up to 6 days postinoculation.

(a) *Materials*   Same as for YS harvest.

(b) *Method*

(i) Open the egg over the artificial air cell.

(ii) Follow procedures for pool harvest (Chapter 6, Section A). Homogenize the tissue as discussed above.

H. EXTENDING THE USE OF EMBRYONATING EGGS IN VIROLOGIC STUDIES

A last comment is in order concerning the importance of eggs in virus propagation. Direct use of the developing embryo has been basic to studies of viruses for a number of years. In addition, the avian embryo is of great importance in supplying material for cell culture preparations (Chapter 3). Before you decide whether to use eggs directly or to use them to produce cell cultures, consider the advantages or disadvantages of both.

Tissue Culture Propagation

THREE

The first successful cultivation of a virus (polio) in tissue culture was accomplished in 1936 by Sabin and Olitsky [44] using human embryonic nervous tissue. It was not until 1949, when Enders, Weller, and Robbins [16] discovered that polio virus could multiply in cultures of non-nervous tissue, that great strides could be taken in research on this, and other, viral agents. It was quickly noted that tissue cultures were less expensive to maintain and required far less room than did live animals. Media requirements for maintenance of the cultures was often a problem but, for the most part, even this has been overcome.

At this point, you are not concerned with the more exotic tissue culture systems. Rather, two standard methods of preparing cell cultures will be discussed. These procedures are applicable to many tissues and if you have success with the examples which follow, you can then proceed to more complicated techniques with some degree of confidence.

A. MATERIALS

For any of the methods described, the following materials should be available. Formulas for preparation of the various fluids, sterilization procedures, etc., may be found in Chapter 9.

> Hanks' growth medium, sterile
> Hanks' maintenance medium, sterile
> Phosphate buffered saline (PBS), sterile
> Trypsin, 0.25% in PBS, sterile
> Commercial serum, sterile

Rappaport stain
Ethyl alcohol, 95%
Disinfectant
Aluminum foil
Sharp scissors
Forceps
Petri dishes, sterile
Pipets, 1.0 and 10.0 ml, sterile
Kimwipes
Gauze pads, 4″ × 4″, sterile
Small funnel, sterile
Centrifuge tubes, sterile
Graduate cylinder, 100 ml, 250 ml, sterile
Erlenmeyer flask, 100 ml, sterile
16 × 125 mm test tubes, sterile
White rubber stoppers, No. 0, sterile
Bunsen burner
Large, flat tray
Small pan
Ice

B. CALF KIDNEY CELLS

Calf kidney cells are relatively easy to work with and you will normally have success in your first attempt to prepare epithelial cell cultures from them. Failure can usually be traced to the fact that the kidneys are not fresh. They should not be obtained from the local supermarket; what are called "fresh" kidneys have often been frozen or canned. Obtain kidneys (from calves under 6 months of age) from a local slaughterhouse. The owners are usually quite willing to give you such kidneys when told what they are to be used for. Ask that the kidneys be removed directly from the carcass with the suet (fat) still in place. Wrap the kidney in foil, place in a chilled, insulated container. At the laboratory, store the kidneys at 4°C until ready to use, preferably within 24 hr.

The trypsinization procedure described is essentially that of Youngner [61] with certain modifications.

1. Materials

General materials, above
Fresh calf kidneys

2. Preliminary Preparation

a. Wash down the work area with disinfectant. Flame a large sheet of aluminum foil and place it, flamed side up, on the work surface.

b. Remove the fat (suet) from the kidney. Grasp the heavy end in the left hand and, with scissors which have been dipped in 70% ethyl alcohol and flamed in the Bunsen burner, cut a slit through to the kidney itself.

> *Note:* Be sure to hold the scissors with the points *down* or else the alcohol may run down over the hand, causing a burn or setting the sleeve of your lab coat afire.

The fat will be quite hard after refrigeration. You can easily peel it away by running the thumb of the left hand down along the slit while at the same time holding back the kidney with the scissors held in the right hand (Fig. 3-1).

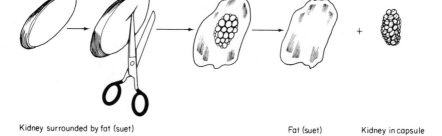

Kidney surrounded by fat (suet)   Fat (suet)   Kidney in capsule

Fig. 3-1. Removing fat from calf kidney.

c. The kidneys are enclosed by a very thin, transparent membrane (the capsule) which must be removed.
   (1) Sterilize the blades of the scissors as described above.
   (2) Hold the scissors at a very shallow angle. Stab the point through the capsule and slide the scissors away from you so that a slit is made in the capsule (Fig. 3-2).
   (3) Grasp the capsule with the left hand and pull it away while simultaneously slitting the capsule with the scissors:

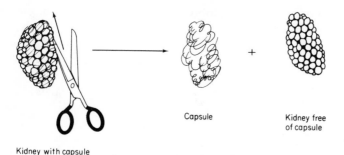

Kidney with capsule   Capsule   Kidney free
                                                       of capsule

Fig. 3-2. Removing capsule from kidney.

When a large enough piece of capsule is loosened, you can usually use both hands to fold the capsule back away from the kidney and, more or less, pop the kidney out of the capsular shell.

d. Resterilize the scissors. Cut through one of the lobes of the kidney. You will see two rather distinct areas, the outer, deep red zone called the *cortex*, and the lighter, inner area called the *medulla*. It is the outer, red area which you want (Fig. 3-3).

Fig. 3-3. X-section of cattle kidney.

    (1) Cut small pieces of the cortex off the lobes with sterile scissors, being careful to remove as little of the medulla as possible. It is not necessary to remove every bit of cortex. In fact, one kidney should suffice for millions of cells.

    (2) Place the small pieces of tissue into a sterile wide-mouthed tube. Continue in this manner until all the lobes have been freed of cortex.

    (3) Resterilize the scissors. Insert the tips directly into the tube, snip the kidney tissue into pieces of approximately 1 mm.

e. Once the tissue has been minced, proceed to the washing step (Section E).

C. AVIAN EMBRYO CELL CULTURES— FIBROBLASTS

1. Materials

General materials, as previously listed
Viable embryonating eggs of 9 days incubation (one embryo will provide enough cells to make approximately 50 ml of cell suspension of the proper concentration)

2. Method

a. Candle eggs to be sure of viability.
b. Disinfect the shell over the air sac. Open the shell by tapping and chipping away with sterile forceps.

   c. Pull aside the membrane with sterile forceps. Lift out the embryo
by grasping the neck lightly between the forceps and clipping
away membranes which attach the embryo to the yolk. Place
the embryo into a sterile Petri dish [Fig. 3-4(a), (b)].

   d. Repeat the procedure until enough embryos are obtained.

   e. Remove the head and internal organs from the embryos; i.e.,
leave only the body for cell culture use. Since the tissue is very
tender at this stage of development, the organs and head can

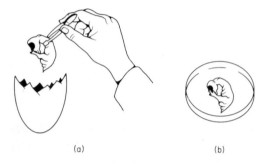

(a)   (b)

Fig. 3-4. (a) Removing embryo from egg. (b) Embryo in Petri dish.

Fig. 3-5. Removing head of embryo.

Fig. 3-6. Cutting up embryo.

be removed simply by grasping them with sterile forceps (Fig. 3-5).

f. Put embryo bodies into a wide-mouthed tube. Cut the embryos into pieces of about 1 to 2 mm (Fig. 3-6). Then proceed to the washing step (Section E).

D. CHICK OR DUCK KIDNEY CELL CULTURES—EPITHELIAL CELLS

1. Materials

General materials, as previously listed
Day old chicks

2. Method

a. Chicks may be anesthetized with ether or chloroform. The authors prefer to grasp the chick in both hands with the thumbs and forefingers supporting the neck. Hold the neck across a sharp edge of a bench or sink and press sharply and firmly with the thumbs. The vertebrae will separate.

b. Completely wet the bodies with disinfectant. This reduces contamination by dust from the feathers and skin. Lay the birds on a clean tray.

c. With the right hand, grasp the ventral skin near the legs and tear the skin foreward toward the head. When the skin has been removed from the abdominal areas of all the chicks, open them as follows.

   (1) Hold the bird with its abdomen facing you by grasping its legs between the index and middle fingers of your left hand. The ring and small fingers should support the back of the bird (Fig. 3-7).

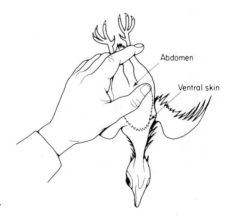

Fig. 3-7. Holding the chick.

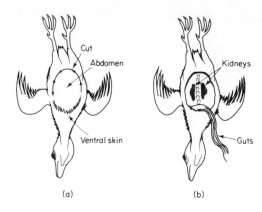

Fig. 3-8. (a) Opening chick. (b) Exposing kidneys.

(2) Still holding the bird by its legs, cut through the body with sterile scissors. Begin at the right side of the abdomen, continue around by the legs, in a semicircle, to the left side of the abdomen. The outer skin, abdominal wall, and guts will drop forward and hang from the bird. The area of the backbone will be exposed. The kidneys lie flat on either side of the backbone, held in place by the fascia (Fig. 3-8).

d. Clip the kidneys loose with sterile scissors and transfer them to a sterile wide-mouthed tube. You can also pick the kidney up by sliding sterile, curved forceps through the fascia and under the kidney. Slide the forceps toward the legs. The kidney should ride up on the forceps and you can lift it out with ease. Place the kidneys in a sterile wide-mouthed tube. With sterile scissors, cut the kidneys into pieces no larger than 1 to 2 mm. Proceed to the washing step (Section E).

E.   WASHING PROCEDURE

1. Materials

PBS, sterile
Scissors, sterile
Small flask, sterile

2. Method

a. Holding sterile scissors almost closed, use them as a spatula to pick up the tissue pieces. Place them in a small flask.

b. Add about 20 ml of PBS, swirl the flask, and allow the larger pieces of tissue to settle. Pour off the cloudy supernate which contains cell debris, blood cells, connective tissue, etc., and discard it.

c. Repeat step b until all the loose pieces of connective tissue, blood, etc., are removed (Fig. 3-9).

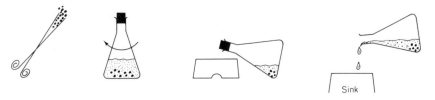

Fig. 3-9. Washing procedure.

d. Once the tissue has been minced, two methods may be used to break the connective tissue and free the individual cells. Continue on to either the "rapid" (Section F) or "overnight" (Section G) methods of trypsinization. The so-called "rapid" method is done on the same day the tissue is cut and an experienced person will be able to accomplish the complete procedure in 3 hr. There are occasions when it is not convenient to begin the trypsinization procedure on that day. You can then use the overnight treatment at cold temperatures and finish the process the following morning. In the latter instance, proceed to the overnight trypsinization method.

F.   RAPID TRYPSINIZATION PROCEDURE

1. Materials

> Trypsin, 0.25%, sterile, prewarmed to 37°C
> Magnetic stirring bar, sterile
> Stirring plate
> Pan with ice

2. Method

a. To the "clean" tissue, add 10 volumes (i.e., roughly 10 times the volume of liquid as you have solid material) of prewarmed (37°C) 0.25% sterile trypsin. Retain the rest of the trypsin at 37°C.

b. Add a sterile magnetic stirring bar. Place the flask on a stirring plate and stir the suspension approximately 10 min at a speed which is relatively rapid but does not cause the formation of foam. Excessive foaming will cause disruption of cells [Fig. 3-10(a)].

c. While waiting, fold the aluminum foil up over the remnants of the kidney and suet (or chick carcasses) and place the package in the incinerator or in the animal room disposal container. Do not throw the package into the waste basket.

d. After 10 min, remove the flask, support it at an angle of 45° and allow the large particles to settle [Fig. 3-10(b)]. Carefully pour off and discard this first supernate (Fig. 3-9), which contains remaining cell debris and a few loose cells from the cut edges of the tissue.

e. Add fresh, warm trypsin. Respin. Allow the cells to settle once again.

f. Pour this supernate, which now has some free cells loosened from the tissue, into a small flask of sterile Hanks' growth medium kept in a container of ice water. This will stop the action of the enzyme [Fig. 3-10(c)].

g. Repeat the addition of trypsin, spinning, and removal of the supernate to the cold flask until there is relatively little tissue left. Discard the remnants. Continue on to the Cell Washing Procedure (Section H).

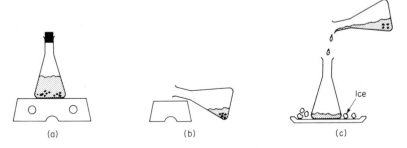

Fig. 3-10. Rapid trypsinization procedure.

G. OVERNIGHT TRYPSINIZATION PROCEDURE

1. Materials

Trypsin, 0.25%, sterile
Prescription bottle, sterile
Black rubber stopper to fit bottle

2. Method

   a. Transfer the "clean" tissue to a sterile prescription bottle. Use one which will contain the tissue pieces in a single layer on the flat side of the bottle.

   b. Add enough cold trypsin to the bottle so that when the bottle is turned on its flat side, a $\frac{1}{4}$-in. deep layer of trypsin will cover the tissue. Experience will tell you how much to use. Stopper the bottle; cover the stopper with foil.

   c. Place the flask on its side in a refrigerator. Be sure that the tissue is well dispersed over the flat surface under the trypsin (Fig. 3-11).

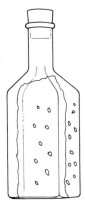

Fig. 3-11. Trypsinization bottle.

   d. In the morning, prewarm a bottle of trypsin to 37°C. Pour the cold trypsin off the tissue. Add a replacement volume of warm trypsin to the bottle.

   e. Manually shake the bottle until the trypsin becomes dark and heavy with suspended tissue cells (Fig. 3-12). Proceed immediately to the Cell Washing Procedure (Section H).

Fig. 3-12. Shaking procedure.

H.   CELL WASHING PROCEDURE

1. Materials

> Centrifuge tubes or bottles, size appropriate for volume
>   of suspended cells, sterile
> Glass funnel, sterile
> 4″ × 4″ gauze squares, sterile
> PBS, sterile
> Hanks' growth medium, sterile

2. Method

a. Remove the aluminum foil from the stem end of a sterile glass funnel. Remove the foil cover from the centrifuge tube or bottle and quickly place the funnel stem into the container.

b. Remove a sterile gauze pad from its wrappings and, handling it by the corners, open it up.

c. Remove the cover from the funnel and quickly place the open gauze square over the funnel opening. Repeat with another gauze square.

d. Add PBS to the flask or bottle of trypsinized cells.

e. Filter the cells through the sterile gauze into the centrifuge containers, distributing the volume among an equal number of containers. This filtration will remove large tissue fragments.

f. Fill the centrifuge containers with sterile PBS, close tightly with their respective aluminum foil covers, and centrifuge at 130 g for 10 minutes.

g. Pour off and discard the supernatant fluid.

h. Resuspend the cells in fresh, sterile PBS by washing cells from one container to the next until all cells are in a single centrifuge tube or bottle.

i. Fill the container with sterile PBS, cover, and centrifuge as above.

j. Continue washing the cells until the supernatant fluid is clear.

k. Resuspend the cells in Hanks' growth medium and continue on to the Cell Counting Procedure (Section I).

I.   CELL COUNTING PROCEDURE

Before tissue cells can be dispensed into test tubes or dishes, they must be diluted to a concentration which will enable them to grow. If the cells are too numerous or too few, a single-layer-thick sheet (monolayer) of cells will not form. In order to make this dilution, you must first know the number of cells you have. Cells may be counted without being stained, using trypan blue, which colors the dead cells while viable cells remain colorless. Most people, however, find that a procedure which stains all viable cells is preferred. The following method will enable you to see the cells more easily. The counting chamber is that used to do blood cell counts.

1. Materials

> Tissue cell suspension
> Hanks' growth medium, sterile
> Rappaport stain
> Test tube, 12 × 72 mm
> Pipets, 1.0 ml, sterile
> Funnel, sterile
> Graduate cylinder, sterile
> Gauze squares, 4″ × 4″, sterile
> Counting chamber, Neubauer
> Test tube rack
> Microscope
> Kimwipe tissues

2. Method

A. PRELIMINARY PREPARATION

(1) Remove the aluminum foil from the stem end of a sterile glass funnel. With the cover still over the large end of the funnel, place the stem into the mouth of a sterile graduate cylinder.

(2) Wrap the aluminum foil cap of the cylinder back over the mouth of the cylinder and the stem of the funnel.

(3) Remove a sterile gauze pad from its wrappings and, handling it by the corners, open it up.

(4) Remove the cover from the funnel and quickly place the open gauze square over the funnel opening. Repeat with three more pieces of sterile gauze so that the funnel has four layers of gauze over the opening.

(5) Pour the cells from the flask, through the sterile gauze, into the graduate cylinder (Fig. 3-13). If the suspension looks very heavy, add some Hanks' growth medium until you can just see through the suspension.

(6) Remove the funnel and cover the cylinder tightly with a sterile aluminum foil cap.

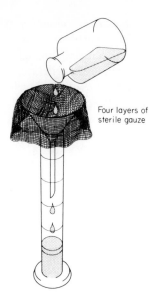

Four layers of
sterile gauze

Fig. 3-13. Filtering cell suspension.

B.   CELL COUNTING PROCEDURE

(1) Place the cover glass over the ruled area of a counting chamber. These have special cover slips which allow correct depth of the chamber beneath. Do *not* use ordinary cover slips for this purpose.

(2) With sterile technique and a sterile 1.0-ml pipet, remove 0.5 ml of well-suspended cells from the graduate cylinder and place in a small test tube.

(3) With a fresh pipet, remove 1.0 ml of Rappaport stain [38] from its bottle, wipe the outside tip of the pipet with a Kimwipe tissue, and add the stain to the cells in the tube.

(4) Mix the contents of the tube thoroughly by gently aspirating with a sterile pipet. Remove a 0.5-ml sample.

(5) Quickly wipe the outside tip end of the pipet with a Kimwipe and place the tip of the pipet to the edge of the cover slip on the counting chamber. Release the pressure *slightly* on the mouth of the pipet, and allow the fluid to run into the counting chamber. It may take a bit of practice to do this. You must allow the fluid to fill only one side of the chamber. Do not add so much fluid that it flows into the channels on each side of the counting area. It might be good practice to try using colored water until you get the technique down to a point where you feel comfortable doing it (Fig. 3-14).

(6) Allow the cells to settle for 2 min. Carefully lift the chamber and place it on the microscope stage.

(7) With the low power objective in place, focus on the ruled area of the chamber. The counting chamber actually has two ruled areas, one

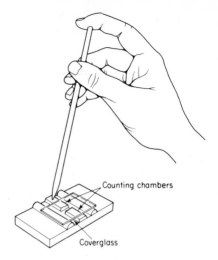

Fig. 3-14. Filling counting chamber.

on each side of a central trough (Fig. 3-15). One ruled area will be found to be sectioned like this:

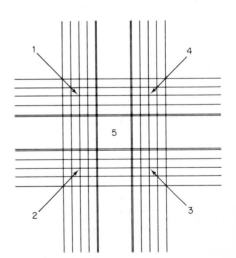

Fig. 3-15. Counting chamber grid.

Focus your attention on the four corner blocks (marked 1, 2, 3, 4, above). Each of these square millimeter areas is divided into 16 squares. With the cover slip in place, the volume over one large square is 0.1 c. mm.

(8) You will be concerned first with finding the average number of cells per milliliter of your concentrate.

   (a) Count the cells in several squares and obtain an average number of cells per square. You may count only the four corner squares on both ruled areas or, to increase your accuracy, count the center squares also.

EXAMPLE

	Square No. 1	2	3	4	5	6	7	8	9	10
Count (No. of cells per square):	36	40	38	41	*	39	43	41	34	*

*Center square of each side not counted.

(b) Determine the average number of cells per square. From the example above where eight squares were counted,

$$\frac{39 \text{ cells per square}}{8\overline{)312}}$$

(c) Adjust for the dilution. In preparing the materials for counting, you combined 0.5 ml of cell concentrate (one volume) with 1.0 ml (twice as much, or two volumes) of stain diluent. You have, therefore, made a $1 + 2$ (or 1 in 3, or 3X) dilution of the concentrate. From step b above, you know you have 39 cells per square of the stain-diluted concentrate. You must now compute how many more cells there would have been had the material been counted undiluted. To do this

Take the average number of cells per square:	39
Times the dilution of concentrate:	3
To get the total number of cells per 0.1 cubic mm (millimeter) of concentrate	117

(d) Now, to correct this value to a count per milliliter of concentrate, you must multiply by 10,000. The rationale is as follows: 1 ml = 1 cc (cubic centimeter).

1 cc is represented by a cube which is 1 cm (or 10 mm) on each edge.

1 cc has $10 \times 10 \times 10$ mm = 1000 cubic mm.

Our count of the average number of cells per square was based on the volume of that square, i.e., 0.1 cubic mm.

To bring this to a 1-cm value, multiply by 10.

To bring the 1-cm value to 1-m*l* value, multiply by 1000 or, combining the two, multiply the average number of cells per square by 10,000. In brief, average number of cells per square $\times$ dilution factor $\times$ 10,000 = number of cells per milliliter of concentrate.

(9) Enumerate all the cells with clear-cut nuclei and surrounding cytoplasm which appear in the white cell (four corner squares) areas:

    (a) Count single cells as one cell [Fig. 3-16(a)].

    (b) Count clumps in which individual nuclei and cytoplasm are
easily visible as clumps of *single* cells, and count *each* cell
[Fig. 3-16(b)].

    (c) When individual cells are not easily discernible as such, clumps
should be counted as a *single* cell [Fig. 3-16(c)].

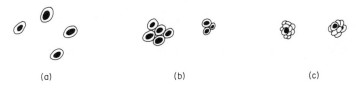

(a)   (b)   (c)

Fig. 3-16. Criteria for cell count.

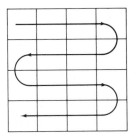

Fig. 3-17. Path of cell counting.

    (d) In counting the cells, develop the technique of counting from
left to right on the first row, right to left on the second row,
left to right on the third row, right to left on the fourth row
(Fig. 3-17). Make it a matter of technique to include in the count
those cells which touch the inner line on the top and right-
hand side of the first and third rows and the top and left-hand
side of the second and fourth rows (Fig. 3-18).

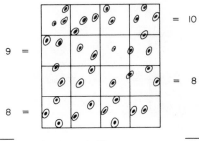

Fig. 3-18. Cell count.

(10) Divide the total number of cells in all four corner squares by four to find the *average number of cells per square*.

(11) The average number of cells per square times 10,000 (correction factor) times 3 (dilution factor) will give the cell count per milliliter, i.e., the number of cells in each milliliter of fluid in the graduate cylinder.

J. DILUTION OF CELL SUSPENSION

Prepare several sets of tubes with cells varying in number from 150,000 per milliliter to 300,000 per milliliter. Individuals will vary in their counting technique, and that number of cells which will produce a good cell sheet for one person will not be the same for another. The type of tissue will also determine the dilution. Cell concentration may vary between 2 and 8×10^5.

1. To determine the dilution factor (i.e., how much to dilute the concentrated cells in order to obtain 150,000 or 300,000 cells per milliliter), divide the cell count per milliliter determined above by the number of cells per milliliter desired.

EXAMPLE

Cells per milliliter of concentrate: 1,000,000
Cells per milliliter desired: 200,000

$$\frac{1,000,000}{200,000} = 5$$

This means that you must dilute the concentrated cells five times in order to

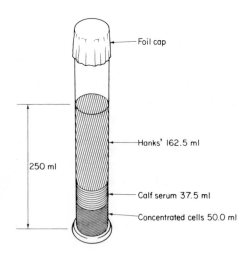

Fig. 3-19. Dilution of cell suspension.

obtain 200,000 cells per milliliter; i.e., you must dilute one part of cell concentrate with four parts of diluent.

Let us say that you have 50 ml of concentrated cells which you have determined must be diluted five times. This means you will have a final volume of 250 ml of cells at a concentration of 200,000 cells per milliliter. However, 10 to 20% of this final volume must be calf serum; therefore, if you use 15% serum, 15% of 250 is 37.5 ml.

2. With sterile technique, add 37.5 ml of calf serum to the graduate cylinder containing the 50 ml of concentrated cells (Fig. 3-19).

3. Add Hanks' growth medium up to the 250-ml mark. Cover with foil.

4. Rinse the cover glass and chamber in running water, wipe dry with soft tissue, and put away.

K. DISPENSING CELLS

1. Preliminary Preparation

With sterile technique, pour the diluted cell suspension into a sterile Erlenmeyer flask containing a magnetic stirring bar. Dispense 1.0 ml of the cell suspension into 16 × 125 mm sterile test tubes or into Petri dishes, glass or plastic. If the number of tubes or dishes is small, you can use a sterile 10-ml pipet, placing 1 ml of suspension in each tube or 6 to 8 ml into each plate. If a large number of tubes or plates are prepared, use a Cornwall syringe as follows.

a. Open the package containing the Cornwall syringe until the rubber tubing holding the steel ball is reached. *Do not touch the rubber tubing.* Lift the tubing by grasping the syringe and tubing near the automatic valve [Fig. 3-20(a), (b)].

b. Carefully remove the aluminum foil cover from the ball. Lift the aluminum cap on the Erlenmeyer flask of cells just enough to permit you to drop the steel ball and rubber tubing into the cell suspension. Press the aluminum foil cap back over the top of the flask and the rubber tubing, but do not depress the rubber tubing [Fig. 3-20(c)].

c. Remove the stopper from an 18-gauge sterile needle container and drop the needle into the Luer-Lok of the Cornwall syringe; twist to lock the needle in place while lifting the container slightly [Fig. 3-20(d)].

d. Fill the barrel of the Cornwall syringe with cells as follows.

(1) Adjust the screw near the handle so that the plunger of the syringe rests at the 1-ml mark. Tighten the arrest screw (the one closest to the metal holder) [Fig. 3-20(e), (f)].

(2) Grasp the rubber tubing (closest to the point where it

Fig. 3-20. Cornwall syringe.

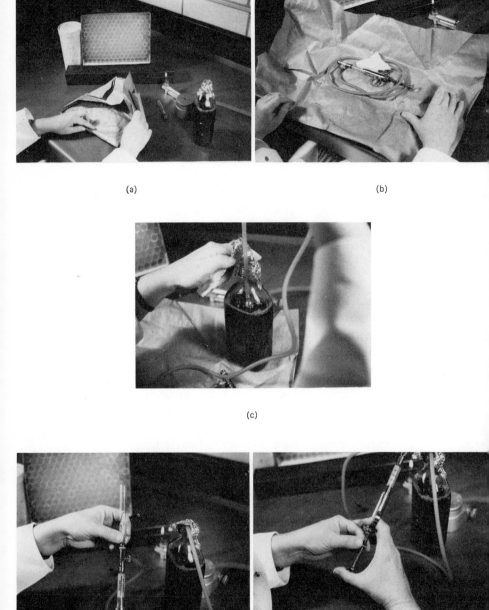

(a)

(b)

(c)

(d)

(e)

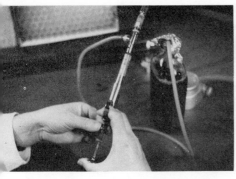

(f)

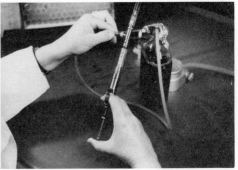

(g)

(h)

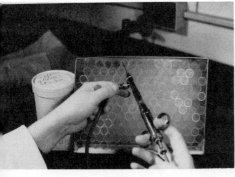

(i)

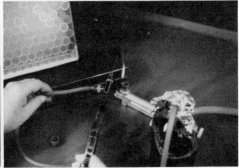

(j)

enters the automatic valve) between the thumb and forefinger of the left hand.
(3) Grasp the metal holder in the right hand with the forefinger
and middle finger through the two metal rings and the
thumb on the metal plunger.
(4) Squeeze the tubing with the left hand [Fig. 3-20(g)].
(5) Push the plunger in with the right hand while releasing the
left hand [Fig. 3-20(h)].
(6) Squeeze the tubing with the left hand while releasing the
plunger in the right hand.
(7) Continue this type of pumping action until the fluid enters
the barrel of the syringe.

2. Filling Test Tubes

a. Set a container of sterile tubes on the counter with the front
edge supported on a pipet canister. Remove the aluminum foil
cover from the box of tubes. Flame the mouth of all the tubes
by quickly passing the Bunsen flame over them.

b. Remove the glass holder from the needle and pump 1 ml of
cells into each tube. *Do not aim at the bottom of the tube.* Pump
the cells into the tube at an angle (i.e., aim for the side of the
tube); otherwise the fluid will hit the bottom and splash back out.
With time and practice, you will develop a great deal of speed
[Fig. 3-20(i)].

c. Flame the needle and replace the glass container over it if you
plan to fill more tubes. Flame the mouths of the test tubes
[Fig. 3-20(j)] once again. Re-cover with flamed foil. Continue to
fill all tubes necessary.

d. Remove the cover from a container of white rubber stoppers
and shake some of the stoppers out into the cover (Fig.
3-21).
(1) Hold a tube in the left hand and pick up a stopper in the
right hand.
(2) Flame the mouth of the tube and the stopper.
(3) Tightly seat the stopper into the mouth of the tube with a
firm twisting motion (Fig. 3-22) and place the tube in a rack,
if you have a small number, or into an empty box having
a 5° front tilt (Fig. 3-23). Mark each tube by running a
marking pen across each row as it is made (Fig. 3-24).
This will indicate the "up" side of the tube and enables you
to tell on which side (the "down" side) the cell sheet has
grown (Fig. 3-25).

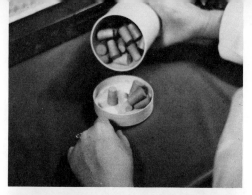

Fig. 3-21. Pouring out stoppers.

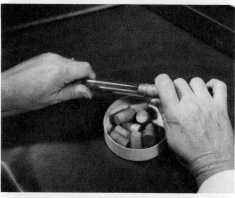

Fig. 3-22. Stoppering tube.

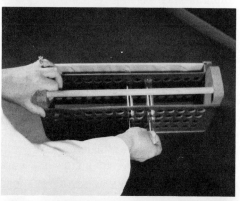

Fig. 3-23. Putting tube in rack.

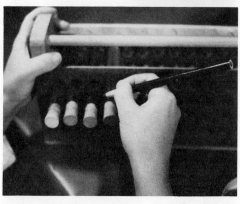

Fig. 3-24. Marking tubes.

Mork up

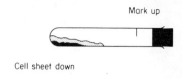

Cell sheet down

Fig. 3-25. Side view of tissue culture tube.

3. Filling Petri Dishes

    a. Arrange Petri dishes, right side up, on a tray.

    b. Carefully lift the lid of each dish just enough to accommodate the needle of the Cornwall syringe. Add a sufficient volume of suspension to the dish. Quickly close the cover.

    c. Because Petri dishes cannot be sealed, they must be placed in an incubator which will maintain a proper balance of 5% CO_2 and humidity. If a CO_2 incubator is not available, lactalbumin hydrolysate medium with Hepes buffer instead of phosphate buffer is recommended [4]. Primary kidney cell cultures can be grown and maintained in ordinary air incubators without CO_2 for periods in excess of 4 weeks if the media is changed weekly. In some cases you may require Hepes buffer in addition to phosphate buffer in the medium.

    d. Place the dishes at 37°C for approximately 4 days.

L. CHANGING THE MEDIA

If the cells have survived, divided, and formed a monolayer, the medium which was originally a cherry red color (pH 7.4 to 7.6) will have changed to yellow (pH 6.8 or less). If the medium has become a definite bluish red, this means the cells have not survived. Presuming, however, that you have been successful, it becomes necessary to change the medium from a growth type to a maintenance type.

1. Materials

        Hanks' maintenance medium, prewarmed to 37°C
        Cornwall syringe, sterile
        18-gauge needle, sterile
        Aluminum foil
        Test tube box
        Disinfectant
        Commercially available serum, sterile

2. Method

a. Wash down the bench with disinfectant. Flame a large piece of aluminum foil and place it, flamed side up, on the work area.

b. Remove the stoppers from the tubes and place them on the aluminum foil. This can be done in a haphazard manner, but the authors find it is more efficient to line the stoppers in rows with the smaller end resting on the foil. Discard the media from the tubes by pouring it into a discard pan.

c. Place the tubes, mouth out, in another box with a 5° tilt. Flame the mouth of the tubes.

d. If plates were used, the medium can be removed with a pipet. Preferably, a sterile Pasteur pipet attached to a suction flask works very well. Remove the cover slightly and tilt the dish with the left hand. Hold the pipet against the lower edge of the dish and suck the fluid away. Be careful that you do not touch the monolayer. Re-cover the dish.

e. With sterile technique, set up a Cornwall syringe and immerse the steel ball in a bottle of Hanks' maintenance medium and fill the syringe. Add 1.0 ml of media to each tube or a replacement volume to each dish.

f. Flame each tube and each rubber stopper; replace the stoppers tightly. The cell sheets can be inoculated immediately or may be used over a period of approximately 2 weeks with an occasional medium change.

g. Place the tubes, marked side up, back in the rack (or box) and reincubate the tubes or dishes of cells.

M. REPASSAGE (TRYPSINIZATION) OF TISSUE CULTURE CELLS

When it is inconvenient or when it is financially impossible to obtain fresh materials each week for making cell culture preparations, it may be necessary for you to keep a particular cell culture going for many weeks. The procedure is simple and easy to perform.

1. Materials

Tissue culture cells
Prescription bottles, sterile
Trypsin, 0.25%, sterile, warmed to 37°C
Pipets, 1.0 and 10.0 ml, sterile
Hanks' growth medium, plus serum, sterile, prewarmed to 37°C

2. Method

a. To the bottle, Petri dish, or test tube of cells from which you have poured off the old medium, add warm 0.25% trypsin. The amount to add will depend on the container used to grow the cells. Use that amount which just covers the cell sheet.

b. Leave the trypsin on the cell sheet for 1 min; then pour it off.

c. Place the container in a 37°C incubator for 15 min. Check the cells under the microscope during this time to be sure that the cells are not peeling off too soon. At this point, you should be able to see that the sheet has holes in it and is beginning to peel from the surface of the glass.

d. Add a small amount of fresh, warm medium and gently shake the container to loosen the cells. You can use a sterile rubber spatula ("policeman") to gently remove the monolayer from the surface of a Petri dish. You can also gently aspirate the cells and medium by drawing the fluid into a pipet, tilt the dish, and allow the suspension to flow down over the surface of the dish again. Pool all the material (if you have several containers of cells) in a flask of cold medium.

e. Do a cell count to determine the number of cells per milliliter of harvested suspension. Dilute the cells to the required concentration with fresh *growth* medium (plus 20% serum, if desired).

f. Redistribute the cells into clean, sterile bottles, dishes, or tubes. Incubate the cells until a new sheet has formed. Remove the growth medium and replace it with maintenance medium.

N. STORAGE OF CELL LINES

In some instances you may wish to maintain a particular cell line (HeLa, Hep-2) for serial culture. These cells should be cooled in stepwise fashion immediately following harvest.

1. Materials

> Cell suspension
> Glycerin, 60%, sterile
> Centrifuge tubes, sterile
> Screw-capped vials, sterile
> Pipets, 10.0 ml, sterile

2. Method

a. Collect the cells by trypsinization as discussed previously.

b. Centrifuge at 205 g for 10 min.

c. Prepare a 1:20 suspension of packed cells in growth medium plus sterile glycerin in a ratio of 0.5:8.5:1. Final cell concentration is 5%.

d. Distribute 1.0-ml aliquants into screw-capped vials, label, and place in a box.

e. Store in a refrigerator at 4°C for 2 hr.

f. Transfer the box to a freezer at −20°C for 2 hr or until the tissue is frozen.

g. Transfer to a deep freeze at −70°C or lower for long-term storage.

h. Specimens to be frozen in liquid nitrogen should be placed in the special vials or ampuls designed to fit the holding rods of the particular tank you have. Follow manufacturer's directions. Add liquid nitrogen periodically to maintain the very low temperature (−196°C).

Propagation and Titration
of Virus

FOUR

A. INTRODUCTION

The correct processing of infected tissues or fluids is necessary if virus is to be isolated and then used to produce a reference supply. In this section will be described the use of those materials which were collected and prepared according to methods outlined in Chapter 1; the isolation, harvest, repassage of virus; and the determination of the number of organisms (titer) in a virus suspension.

1. Preliminary Passage

The original material (O.M.), which may or may not contain virus(es), must first be inoculated into a susceptible host (animals, eggs, cell cultures). This is usually accomplished by preparing suspensions of infected tissues (Chapter 1) for inoculation into an "indicator system" which is used to "indicate" the presence of virus by some measurable response. In order to assure that any observable changes are, in fact, due to the inoculum and not some agent innate to the indicator system itself, a proportionate number of indicators are reserved as uninoculated controls. Death, abnormalities, paralysis, cytopathic effects, etc., in this indicator system will be your first clue to the probable presence of virus. Since viruses do not always multiply to sufficient numbers in the few days that some systems can be maintained (and you will not, therefore, see an indication of viral presence), it is common practice to collect these tissues and repass them (*blind passage*).

Each time a virus is passed, the possibility increases that it may be altered in some way by the process. In addition, there is the danger of inadvertently using animals, eggs, or cell cultures which may have become infected with other viruses which have not yet multiplied to sufficient numbers to make their presence known. An even more difficult aspect of viral propagation is the problem of hidden (*latent*) viruses which may be present in many living systems. Such viruses do not express themselves until metabolic change occurs in the systems which harbor them, initiating growth and multiplication of these hidden agents. If such a situation occurs in a system you are using, the virus which you are studying and these latent agents may thus be mixed together. A careful characterization of the virus from the original material, and a similar characterization of the virus you are propagating and testing, will permit comparison. Any differences in characteristics must be carefully checked.

2. Harvest

The inoculated host material, which may or may not indicate the presence of virus, must be collected (*harvested*). This harvest is labeled and stored as "first passage" material and may be used in many tests. This prevents the exhaustion of your original reference sample, some of which must *always* remain available to others who may wish to repeat (and, sometimes, to challenge) your work.

The development of a supply of materials usually proceeds as shown in Fig. 4-1.

Note that a certain proportion of the first passage harvest is reserved and stored in small aliquants for the preparation of second passage material. This extends the time period over which you may work with harvests only once removed from the original. In some laboratories, studies are conducted only on second passage harvests, the first passage materials being used to make almost endless supplies of second passage virus-infected samples. Only where confirmation of the test results are desired are first passage and original materials then used.

3. Repassage

Hopefully, there will be enough viruses in the first passage harvest to infect more of the cells (or animals or eggs) in the next passage than was possible with the O.M. put in the indicator system the first time. The purposes of repassage are many. Initially, you are attempting to determine the presence of virus, and you may have to repass the material several times depending on the virus, indicator system, time, expense, etc. If first passage material is positive, you must repass it for the purpose of obtaining a larger volume

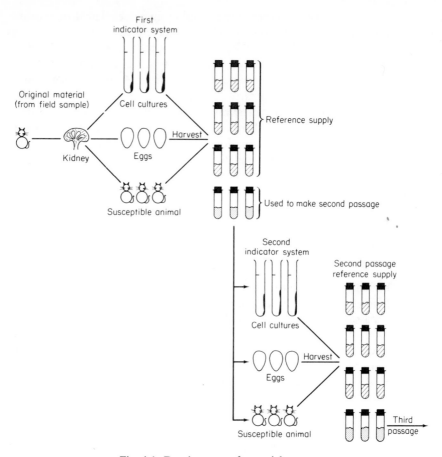

Fig. 4-1. Development of materials.

of the virus for study. You must repass in order to build up the virus titer and you must repass the virus in order to study its biochemical and biophysical characteristics. If the original passage is negative, you should repass at least twice more before you decide no virus is present. Normally, three repasses are recommended. Even then you may decide on several more passages or you may decide to try a different indicator system.

Repassage of viruses in cell cultures can present problems. Although these cells grown in tubes may have come from the same species of animal from which the virus was originally isolated, these cells are being maintained under artificial conditions. This cannot help but alter their function somewhat. It may take several passages before a virus is adapted to cells which are similar to, but no longer the same as, the cells in which it was "growing" originally. It may be necessary, therefore, to repass the virus several times before there are sufficient numbers produced in each infected cell. The question usually arises as to why you don't simply leave the cells unharvested until enough viruses have "grown." The limiting factor here is the metabolic state of the cells in culture. There is a point beyond which they cannot be maintained in a healthy state under artificial conditions. Since viruses must have living cells in which to multiply, the cells and the viruses in them must be harvested before cell death occurs. Dying cells will release their viruses, but if there are no other healthy cells in the culture which the released viruses can penetrate, the latter will die also. The only solution is to harvest the infected cells before they enter a phase of metabolic decline.

Propagation of viruses in eggs involves a somewhat similar problem. Here you are limited by the developmental period of the embryo (e.g., chicken, 21 days) within a confined space (shell). Consequently, the availability of a particular tissue or fluid changes rapidly. You must harvest from, and repass into, embryos at particular stages of development.

It must also be realized that some viruses can be passed in primary cell cultures (i.e., cells prepared directly from tissues), whereas other viruses must be passed in secondary cultures (i.e., cells prepared from previously cultured cells). Some viruses will multiply in several cell lines while others are quite specific for a cell type.

It must be obvious to you that the variables in the passage of virus materials are many. At first glance, they sometimes appear insurmountable. Fortunately, most viruses will propagate in systems the same as, or closely related to, those from which they were initially taken.

Vials containing repassaged materials should be clearly marked, indicating the virus, type of host, and the number of times passed in that host. For example: "ECBO, #48, MK_3, CaK_{10}" indicates that enteric cytopathogenic bovine orphan virus, No. 48, has been passed 3 times in monkey kidney, followed by 10 consecutive passages in calf kidney cell cultures.

4. Isolation

Once you are confident that you have a virus, have repassed it, and have a pool of it in storage, you may proceed in two directions. If you suspect that more than one virus is present in the harvested suspension, you will want to try to separate them under an agar overlay (see Plaque Technique, Chapter 6). If you are then confident that only one virus is present, or if only one virus was present initially, it may be said that you have *isolated* a virus.

5. Titration

You can now determine how much virus is present per unit volume of suspension. A thorough understanding of logarithms is necessary. The calculations of virus titer are essential for almost all subsequent work.

B. PROPAGATION OF VIRUSES

1. Animals

A. INTRODUCTION

Where animals are to be used as the indicator system to propagate and study viruses, certain general points should be considered. Where possible, use *pathogen-free* animals. These are reared in isolation, free from exposure to infectious agents. Such animals have a high degree of susceptibility to those agents which would normally infect them. They lack antibodies to such agents so that it is possible to show a specific increase in antibody level when inoculation or exposure under laboratory conditions does occur. Some animals are reared in what is called *germ-free* environments. These are obtained under sterile environmental conditions by Caesarian section, fed sterilized foods, and maintained in isolation units under controlled atmospheric conditions. Pathogen-free animals are not the same as those called germ-free. In the former instance, the animals are isolated to minimize their exposure to infectious agents. In the latter instance, they are obtained and housed to avoid contact with all viruses as well as bacteria, even if the latter are normal and necessary to the animal's function; i.e., *Escherichia coli*, a common intestinal bacterium, is necessary for the production of vitamin K. Animals lacking or deprived of normal levels of *Escherichia coli* (high levels of some antibiotics will do this) may develop diarrhea and eventually die.

A blood sample must be obtained from the animal to be used in isolating a virus. The serum or plasma must be tested to determine the degree of previous exposure to various pathogens. There are a number of tests which will reveal if the selected animal has been exposed to the pathogen to be studied (see Chapter 5). If these tests show that the animal serum neutralizes or reacts with a sample of the virus, then another animal must be selected. Attempts to propagate a virus in an animal having antibodies to the virus will, obviously, be unsuccessful.

Age is a factor to consider in the selection of an experimental animal. Young animals are generally preferred since their exposure to infectious agents will be less (and, thus, their chances of developing immunity will be greater). Consideration must, however, be given to the fact that very young animals may still have antibodies passed on from an immune parent (passive immunity), and some period of time must pass before these circulating antibodies are eliminated and the animal is susceptible. These periods may range from a few days to several weeks. This decline in passive antibody can be determined by the several serologic methods described in Chapter 5. Ideally, an inoculation is delayed until these antibodies are at a low level.

The matter of immunologic incompetence must also be considered. This refers to the inability of the experimental animal to "recognize" foreign protein and produce antibodies to it. If very young animals are inoculated with overwhelming concentrations of virus, their immunologic (reticuloendothelial) system may be so paralyzed that they cannot respond by producing antibodies. Animals within a few days of birth are most susceptible to this type of immune paralysis. Inocula, therefore, should be administered in small doses, and in dilutions ranging from at least 1 part in 10 to 1 part in 100 of diluent. Several animals should be inoculated with each level if the danger of immune paralysis is imminent.

B. RESTRAINT

Be considerate. It is not necessary to tie an animal so tightly that blood circulation is reduced. It is surprising how many animals will permit inoculations with no, or little, restraint. Rather than risk a broken needle because an animal jumps at an inopportune moment, some restraint is usually necessary. Where animals are secured in unnatural positions, i.e., on their side or on a slant board, be aware that suffocation may occur or circulation may be impaired. This is due to the weight of internal organs pressing against the diaphragm or large blood vessels. The animal's internal supporting tissues and membranes may be unable to counter such unnatural shifts in the weight of his internal organs. Also, in animals whose natural stance is curve-backed (rabbit) or whose spine lacks the lumbar curve of man, securing them so their backs are against a straight board should be avoided. It equals stretching a human on a curved surface with his back pressed inward. A padded and concave securing board will make such an animal more comfortable and reduce the danger of unexpected jumps and twists.

C. INOCULUM

Too high a concentration of virus may be so damaging to the experimental animal that death due to massive infection may occur, but little proliferation of a fresh supply of virus will be produced. The concentration and volume of the inoculum, therefore, should be adjusted to infect the animal but permit its survival for a time sufficient to have the introduced virus multiply, spread to yet uninfected cells, and be released. Sometimes several animals are inoculated, each with a different dilution. In this way, the inoculum, concentration, and duration of infection can be better determined.

D. ROUTE OF INOCULATION

Some consideration should be given to the "natural" route of infection. Most infectious agents enter a susceptible host by inhalation and/or ingestion.

Some enter through breaks in the skin. The target organs may be any of the body systems of the animal: bone, liver, central nervous system, muscles, etc. When inoculating an experimental animal, the choice must be made between using a natural route or inoculating the material in such a way that the target organ will be reached more directly, i.e., by intravenous, intradermal, or intraperitoneal routes.

(1) *Natural Routes*

(a) *Ingestion*   Several obvious methods can be used where virus is to be administered by ingestion. The virus-infected material can be made up in a sweet-tasting liquid. Most experimental animals will willingly drink the suspension. Where direct administration is required, a small diameter tube may be used. Care should be taken to avoid irritating the animal's throat during insertion of the tube. The virus suspension can then by injected into the tube with a needle and syringe. Mixing the virus with food or water is an acceptable method, providing the ability of the virus to survive has been assured. In either instance, withholding food and water for a 24-hr period prior to its being offered for consumption, and the mixing of the virus with a minimum amount of food or water, will increase the chances of the maximum amount of virus being ingested.

(b) *Inhalation*   Virus suspensions may be given by aerosol sprays administered in a closed cabinet. Again, computation of the actual numbers of virus particles in the suspension must be made. The number of viruses inhaled by the experimental animal can only be estimated. Records of such administration should include information on the aerosol equipment used, nozzle size, particles produced per cubic volume, volume of material used, and duration of exposure to aerosol. Since a natural route is being used, some consideration should be given to the approximation of natural conditions. There is no need to administer the aerosol at such high velocities that the animal limits his breathing because of "wind" velocity, nor so long a period of time and large a volume of spray that the animal's skin, fur, or feathers are saturated. A gentle, short burst, repeated a number of times, is usually sufficient. Administration of infectious materials may also be achieved by a tube gently inserted as a bypass of the glottis, with the suspension then inoculated into the tube. This method makes possible the administration of a measured amount and is used more often than aerosol sprays for this very reason. It must be obvious that infectious agents dispersed by the aerosol method will remain suspended in the air for considerably long lengths of time. You should take all necessary precautions to protect yourself and anyone else when working with such aerosols.

2) *Experimental Routes*

(a) *Materials*   All equipment used must, obviously, be sterile, and all procedures must also be done with utmost care for the safety of the personnel involved and the animals to be treated.

(i) *Syringes* Select a size to accommodate the inoculum. If a number of animals are to be inoculated, each syringe should be of sufficient capacity to hold an amount which will inoculate several animals.

(ii) *Needles* Select a needle of as fine a gauge (hole diameter) as possible. Usually there is some parallel between the capacity of the syringe and the diameter of the needle. Although needles fit many different syringes, it makes little sense to select a needle for the administration of 0.1-ml amounts per animal and attach it to a syringe of 25-ml capacity. The pressure created in depressing the plunger of the syringe may blow the needle from the tip. Trial with a water-filled syringe and various needles will usually be a help if there has been no previous experience with handling either. It is better to use small capacity syringes and fill several for inoculation (or refill one) than to use one which is too large. In refilling a syringe, it is best to remove the needle and replace it with a sterile one for each animal or group of animals. Be sure the needle is firmly seated and there is no leakage. Make it a habit to position the needle such that when locked in place, the bevel (hole) is uppermost and in line with the gradations on the barrel. In this way you are assured of simultaneous control over the flow and the amount of inoculum (Fig. 4-2).

(iii) *Disinfection* Frequently, 70% alcohol is used to disinfect the area where the inoculation is to be made. Tincture of iodine may also be applied with sterile cotton swabs.

(iv) *Inoculum* Generally the inoculum is maintained in an ice bath until a short time before use. It should then be allowed to come to room temperature. This is especially important in the inoculation of small animals. An intravenous inoculation of 0.2 ml of ice-cold suspension into a mouse may provoke shock. That same amount inoculated into a dog would be less dangerous. In no instance is it necessary to use ice-cold materials if reasonable care is given to timing the removal of the inoculum from the ice bath.

When preparing the inoculum, be sure that solid particles do not enter the syringe. Such particles may block the syringe and interfere with an injection. They may be of no concern in subcutaneous injections, but particles injected intravenously may cause an embolism.

Be sure that all air bubbles are removed. Fill the syringe, then hold it vertically with the needle pointing upward. Hold the thumb of the right hand against the plunger. Place an alcohol-soaked piece of cotton over the needle. Tap the barrel with fingers to dislodge trapped air bubbles. This will cause them to rise to the top of the liquid and is particularly necessary with plastic syringes (Fig. 4-3). Press the plunger upward until all air is expelled from the

Bevel

Fig. 4-2. Syringe with needle, bevel up.

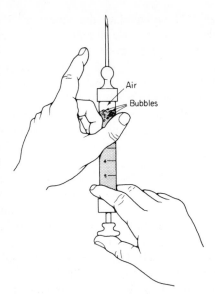

Air

Bubbles

Fig. 4-3. Tap syringe to remove air bubble.

syringe and needle. Any inoculum will be collected in the cotton, which should be burned immediately or placed in a contamination pan.

A large volume is unnatural and, in the case of a subcutaneous injection, for instance, a large bleb will result at the area of inoculation. Any volume should be injected slowly, particularly when intravenous inoculations are being made.

(b) *Methods* The number of animals you will use will depend on cost, housing, etc. If the animals are small (mice, chickens) you will, no doubt, use greater numbers than if rabbits, calves, etc., were being used. The greater the number of animals used for each dilution of virus, the greater your statistical accuracy will be.

The size of the animal and the number per dilution of virus will determine the housing. Animals receiving the same dilution of virus should be housed together where size permits. Prepare the housing units well ahead of inoculation day. Attach record cards to each unit. Add feed and water at the last possible moment.

If cleaning of the animal skin is difficult due to a thick layer of feathers or fur, it may be necessary to remove this from the inoculation area by plucking, shaving, or clipping. As this causes some degree of irritation, try to do it 24 hr previous to inoculation and place the animals in their respective units.

Set out a tray containing 70% alcohol, tincture of iodine, sterile swabs, sterile cotton or gauze pads, sterile needles and syringes of appropriate size, and any other materials you will need except the virus and diluent. Inocula-

tions may be started early on the following day as soon as you have thawed
and diluted your virus sample.

(i) *Intramuscular* Select one of the large body muscles. In most
species, the thigh, lateral thoracic, or abdominal muscles are very convenient.
Disinfect the skin. Steady the skin area between your left thumb and forefinger. Insert the needle with a quick jab. This is most easily accomplished
by holding the syringe as you would when throwing a dart. Slowly express
the inoculum. Quickly remove the needle. Wipe the area with alcohol (Fig.
4-4).

(ii) *Intravenous* Whatever vein is selected for the inoculation,
care should be exercised to assure the inoculation is made *with* (not against)
the flow of blood. Where the direction of flow is not known, it can usually
be determined by watching for the direction of reduced flow when a finger
is pressed firmly in the center of a surface vessel. On one side or the other of
this pressure point, the vessel will blanch due to the lack of blood. This
indicates that the blood must be coming *from* the side filled with blood and
traveling *toward* the side now empty. The needle should be inserted so the
inoculum will be swept along with the blood.

Rabbits are usually inoculated in the marginal vein of one ear and bled
from the opposite ear. Commence at or near the tip of the ear since, with
repeated inoculation, the vein may become occluded. Clip or pluck the
hair. Wash the ear with a mixture of alcohol and xylol to render the vein more
conspicuous. Stand in front of the animal and stretch the ear forward. Grasp
the ear in the left hand with the middle finger under the site of injection and
the thumb and forefinger on top of the ear. In this way, the site is steadied.
Hold the syringe parallel to the ear and thrust the needle (26 gauge) through
the skin in the area held between the thumb and forefinger. Continue through
the wall of the vein into the lumen. If the needle was inserted at a very shallow
angle, it should be in the vein and the blood will be seen to disappear as the
fluid is injected. If fluid is being forced into the subcutaneous tissue instead,
a swelling will occur immediately. Stop at once. Make another attempt farther
down along the vein. Withdraw the needle and place an alcohol-moistened

Fig. 4-4. Intramuscular injection.

cotton or gauze pad over the puncture until bleeding stops. Wash the ear thoroughly to remove the xylol which otherwise will be extremely irritating and will cause low-grade inflammation (Fig. 4-5).

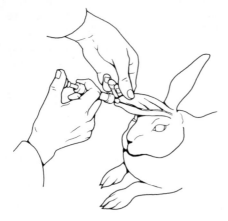

Fig. 4-5. Intravenous injection, rabbit.

Rats may be inoculated in an external jugular vein, mice in a lateral tail vein, and guinea pigs in a large superficial vein on the dorsal, inner side of the hind leg or in an external jugular vein.

(iii) *Intraperitoneal* Care must be taken in doing an intraperitoneal inoculation in order to avoid puncturing the intestine. The easiest area to use for such an inoculation is the ventral side of the body, about 2 in. below and to the left or right of the navel. It is best to raise the animal's hind quarters so that the internal organs slide forward. Small animals (mice, chickens) may be held upside down by their legs. Larger ones (rabbits, cats, dogs) may be restrained for a short time on a slanted (head-downward position) restraining board. Clip the hair and shave the area. Apply tincture of iodine. Lift the skin and underlying connective tissue. Insert the needle for a short distance in the direction of the head. Then hold the needle at a right angle to the peritoneal wall and thrust it through into the cavity. When the peritoneum has been entered, a relaxation of the abdominal muscles will occur. Withdraw the needle slightly, and make the inoculation. Withdraw the needle quickly and swab the point of inoculation with alcohol (Fig. 4-6).

(iv) *Subcutaneous* Shave and disinfect the area. Pinch up the skin between the index finger and thumb of the left hand. Insert the needle. The injected material (amount will depend on the animal) will form a blister or bleb.

(v) *Intracutaneous* Shave and disinfect the area. Pinch up the skin. Insert the needle, lumen up, as superficially as possible. A raised white spot which shows the pits of the hair follicles can easily be seen. Inoculate no more than 0.2 ml of virus.

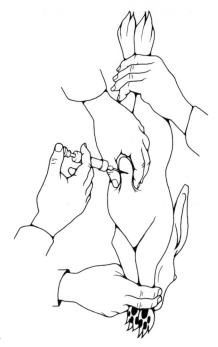

Fig. 4-6. Intraperitoneal injection.

E. RECORDS

It should be obvious that daily records on the inoculated animals should be kept. The information on such records are often determined by the type of animal and purpose of the experiment. The diversity of the points on which records are kept will be appreciated from the following list, which certainly does not represent the gamut of possibilities: weight, food and water consumption (weight and volume measurements), temperature, behavioral patterns, breeding records, bleeding schedule, inoculation schedule, signs of disease (including paralysis, blindness, deaths), etc. It is good practice to keep a 4 × 6 in. white file card in a holder on each cage where this information can be placed immediately. A duplicate set of notes should be placed in your experimental notebook. At the completion of your experiment, the cards may be filed for further reference.

F. HARVEST AND REPASSAGE OF TISSUES

It is often necessary or desirable to necropsy your test animals to obtain samples for future work, to take cultures, or to study gross and microscopic pathology. Necropsies may be performed at any time. It is often wise to collect specimens at several intervals after inoculation since there may be no overt signs of disease. If signs indicate the presence of a pathogen, obviously you will sacrifice the animal and necropsy it. If death occurs, you should perform necropsy as soon as possible.

The tissue(s) to be collected or examined will depend on the particular disease or symptoms exhibited by the animal. Consult various anatomy texts if you are unfamiliar with the particular animal you are using. Follow the methods discussed in Chapter 1 for collection and handling of specimens.

Several precautions should be observed. Where samples are to be taken for future culture, sterile techniques must be observed. Employ sterile instruments at each stage and, after necropsy, place each in a contamination pan immediately. Use rubber gloves, sterile gowns, and masks (if required) where highly contagious diseases are suspected or present. The animal should be disposed of by incineration if possible. Scrub the work area with a 1 % solution of crude cresol.

Label and freeze all samples collected. These may be used for repassage in similar, or completely different, indicator systems if desired.

2. Cell Cultures

A.   INTRODUCTION

In the previous section, the propagation of viral agents in animals was described. For many, the use of animals is not convenient since the cost of buying, housing, feeding, and cleaning them can be prohibitive. Like the situation in animals, viruses have been isolated from cell cultures as well as propagated and titrated in them for further study. This is not surprising since the cells in culture are derived from some living host in the first place. Thus, the *intact*, living host is not the only source of, or indicator system for, viral agents.

B.   MATERIALS

Virus suspension (from infected animal tissue, eggs, or
    other cell cultures)
Hanks' medium or other diluent, sterile
Tubes, plates, or bottles of tissue culture cells,
    appropriate numbers as required
Pipets, 1.0 ml, sterile
Tissue culture racks or flat trays
Bunsen burner
Disinfectant
Paper toweling

C. METHODS

(1) *Test Tube*
  (a) Swab the work area with disinfectant.
  (b) Rapidly thaw a virus sample. Wipe the vial dry with paper toweling. Dilute the virus 10^{-1} in Hanks' medium.
  (c) With sterile technique, remove an aliquot of fluid from the vial. Add 0.1 ml to each of several tubes of tissue culture.
  (d) Place the tubes in a rack, marked side up, along with tubes of uninoculated tissue cultures as controls. Incubate at 37°C.
  (e) Examine for changes (compared with controls) every 1 or 2 days for a total of 7 to 10 days. Use a standard laboratory microscope with low power objective. There are commercially available observation racks which hold the tube in place on the stage. These are merely two parallel pieces of metal (or plastic or wood) attached to a support on each end [Fig. 4-7(a)]. The support at one end should be slightly thicker so that the whole effect is to give a 5° angle from the horizontal [Fig. 4-7(b)]. The unit, which looks like a small piece of railroad track, supports the test tube at enough of an angle to keep the liquid away from the stopper. The track is placed over the light source where it is held by the mechanical stage.
    (i) Place the test tube on the track with the cell sheet *up*, closest to the objective lens and with the top of the tube to the left at the higher (raised) end of the track.

Top View

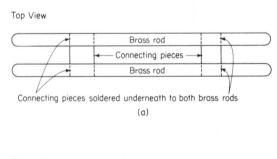

Connecting pieces soldered underneath to both brass rods

(a)

Side View

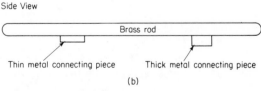

(b)

Fig. 4-7. Tube support. (a) Top. (b) Side.

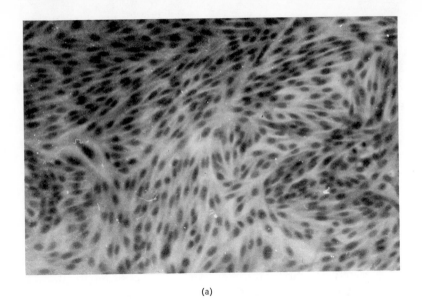

(a)

(b)

Fig. 4-8. Cell culture changes.

    (ii) Move the test tube, *not* the track, with the left hand while focusing up and down with the right hand, using the course adjustment screw only.

Cells may show rounding and clumping, granularity, a foamy appearance, thinning and tearing of the cell sheet, thickening, or no change what-

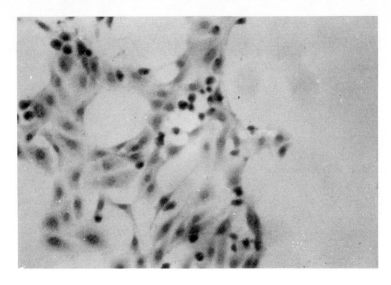

(c)

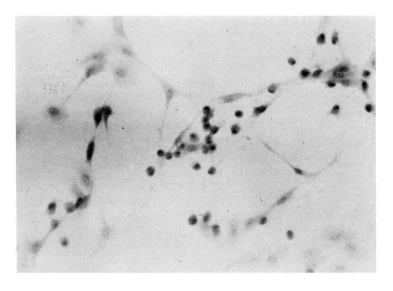

(d)

ever [Figs. 4-8(a)–(d)]. These various cytopathic effects (CPE) are character-istic for each virus and should be noted and recorded.

(2) *Bottles or Petri Dishes*
    (a) With sterile techniques, inoculate bottles or dishes of cells with 0.1 ml of virus suspension. Inoculate controls with 0.1 ml of PBS.

(b) Place containers on a flat tray and manually rotate them for 5 min. Do not be vigorous, as you do not want to splash the liquid over the edge of the Petri dish. Use a gentle motion, first clockwise, then counterclockwise, then forward and back, and then side to side.

(c) Incubate bottle cultures in a standard 37°C incubator. Incubate Petri dish cultures in a CO_2 incubator at 37°C or try Hepes buffer and use a standard incubator.

(d) Examine for change (compared with controls) daily for a total of 7 to 10 days, if possible. Flat-sided bottles and Petri dishes may be placed directly on the microscope stage. The cover should not be removed from the dish. Manually move the dish or bottle around the stage as you observe the sheet through the low power objective using the course adjustment screw.

D. HARVEST

(1) *MATERIALS*

Infected tissue culture cells
Pipets, 10.0 ml, sterile
Rubber policeman, sterile
Vials, screw-capped, sterile

(2) *METHODS*

(a) *Viruses Unaffected By Cold (Table 1-1)*

(i) When half the cell population is found to be affected or when cultures suspected of containing virus are to be harvested, put the tubes, bottles, or dishes in the freezer along with the controls.

(ii) When the material is "slushy," gently shake the containers to loosen cells from the glass. Cultures in bottles or dishes may be harvested by gently scraping the "slush" with a sterile rubber policeman. Remove the suspended cells with a sterile pipet, pool, and dispense into vials. Fluids in tubes may be pooled in one tube using sterile technique and then dispensed into vials with a sterile pipet. Label, date, and freeze.

(b) *Viruses Affected By Cold (Table 1-1)*

(i) Viruses which are known or suspected to be sensitive to freezer temperatures may be collected by gently scraping the cell sheet loose with a sterile rubber policeman.

(ii) Remove the cell-containing fluid with a sterile pipet. Dispense into vials. Label, date, and store at a temperature best suited to the virus.

E. REPASSAGE

Cultures which show CPE on first passage and those which do not (*blind passage*) should be subpassed three times in that cell type. The control material *must* be passed at the same time.

(1) *MATERIALS*

> Infected tissue culture material, plus controls,
>   frozen or freshly harvested
> Uninoculated tissue cultures
> Pipets, 1.0 ml, sterile
> Tissue culture rack or tray

(2) *METHOD*
  (a) Disinfect the work area.
  (b) Rapidly thaw the frozen samples. Disinfect the outside of the vials by wiping with disinfectant-wetted paper toweling.
  (c) Remove aliquots of cell suspensions of virus using sterile techniques. Inoculate 0.1 ml of infected material into each of three tubes, dishes, or bottles of tissue culture.
  (d) Inoculate 0.1 ml of the controls into each of three tubes, dishes, or bottles of tissue culture.
  (e) Replace in the rack or tray along with additional uninoculated control cells. Label and incubate at 37°C in the appropriate incubator.
  (f) Observe for 7 to 10 days and harvest as described previously, above. If *no* CPE is noted in the test inoculates after three subpassages, consider the material to be virus-free. You must decide for yourself, however, whether you wish to proceed with additional passages since some viruses do require longer adaptation time. If CPE is noted in the test inoculates and *not* in the inoculated control cells, and *not* in the uninoculated control cells, harvest the test inoculates as well as the control cells as described above, label, and freeze. If CPE is noted in both the test *and* the control inoculates, this means the tissue culture cells themselves also contain a viral (*latent*) agent, and all of these cultures should be discarded.

3. Embryonating Eggs

See Chapter 2 for the inoculation and harvesting of egg materials. Repassage will follow the same procedures as used for initial isolation.

C. TITRATION OF VIRUS

A virus titration is a quantitative determination of viral activity, i.e., the smallest amount of viral suspension which can produce disease, lesions, or some recognizable effect in the host. The smallest amount of virus which produces this reaction is termed the *infectious unit*, and the *titer* is expressed as the number of infectious units per unit volume.

First you must prepare serial dilutions of the virus. The dilution may be by factors of 10, 5, or 2; the smaller the factor, the more precise the titer. However, the cost, availability, and housing of your indicator system will commonly dictate the dilution to be used. Mistakes are commonly made in the preparations of these serial dilutions so we shall consider this first.

1. Preparation of Dilutions

See Appendix A.

2. Preparation of Serial Dilutions

A. MATERIALS

> Hanks' medium, sterile, prewarmed
> Virus suspension
> Small test tubes, 12 × 72 mm, sterile
> Pipets, 1.0 and 10.0 ml, sterile
> Rack
> Bunsen burner

B. METHOD

(1) Set up a row of 10 small, sterile tubes in a rack. Add Hanks' medium to each tube with sterile techniques.

The quantity used will depend on how much you will need; i.e., if you want to add 0.1 ml of each dilution to each of three tissue culture tubes or eggs or animals, you will need a total of 0.3 ml plus enough to make the next dilution. If you want 0.2 ml of each dilution for each of five tissue culture tubes or eggs, you will need 1.0 ml plus enough to make the next dilution, etc.

In virology, you normally begin dilutions at 1:10 and proceed in tenfold steps: 1:10 ... 1:100 ... 1:1000 ... etc. Therefore, for the first dilution you can use

$$0.1 \text{ ml virus} + 0.9 \text{ ml diluent} = 1:10$$
$$0.4 \text{ ml virus} + 3.6 \text{ ml diluent} = 1:10$$
$$1.0 \text{ ml virus} + 9.0 \text{ ml diluent} = 1:10$$

Let us use the first by way of example, so, add 0.9 ml of Hanks' medium to each tube.

(2) With a *new* 1.0-ml pipet, add 0.1 ml of virus suspension to tube No. 1.

Do *not* insert the pipet into the tube or liquid as the outside of the pipet will have extremely large numbers of virus on it. Just touch the upper rim of the tube with the pipet and quickly allow the suspension to run in. *Discard this pipet into the pipet jar.*

(3) With a *new* 1.0-ml pipet, mix the contents of this first tube.

The technique of proper mixing is important and gives the impression of playing a slide trombone. With practice (you can use dye in water) you can develop this technique to a fine degree.

(a) Hold the test tube in the left hand and pipet in the right. Grasp the cotton stopper in the bend of the little finger of the right hand. Flame the lip of the tube.

(b) Place the pipet in the liquid as far as possible without displacing the liquid from the tube and simultaneously draw the liquid up the pipet.

(c) Expel the contents of the pipet back into the tube while simultaneously lowering the tube to its original position.

(d) Raise the tube to its original position and simultaneously draw the liquid up into the pipet again.

In these maneuvers, the pipet should *not* leave the inside of the tube. The actual distance that you are drawing the tube away is really very slight. The idea is to get plenty of stirring action.

(e) Repeat this "trombone playing" 9 or 10 times.

(4) Remove 0.1 ml of 10^{-1} dilution, flame the lip of the tube, and replace the plug. With sterile technique, expel the 0.1 ml into tube No. 2 without touching the inside of the tube or placing the pipet into the liquid. *Discard this pipet.*

(5) With a *new* pipet, mix the contents of tube No. 2 as in step 3. Continue in this manner until you have removed and discarded 0.1 ml of solution from tube No. 10 (Fig. 4-9).

You will notice that a *new* pipet is used after *each* transfer. In all titrations *a different pipet must be used for each dilution* because a pipet will contain billions of virus particles (or bacteria or whatever) on its outer surface which cannot be rinsed free. If you do not change pipets, you carry these particles along, with the result that you will have a large dilution error. This is probably the one precaution which most fail to observe, and this error is often referred to by the German term *Pipetenfehler.*

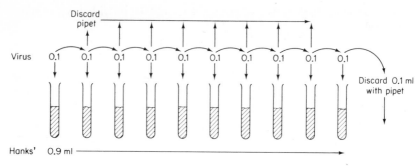

Fig. 4-9. Serial dilution.

3. Titration

A.   MATERIALS

Virus suspension
Uninoculated tissue cultures in tubes, bottles, or dishes;
   embryonating eggs; or appropriate animals
Hanks' medium or other diluent, sterile
Pipets, 1.0 and 10.0 ml, sterile
Test tubes, 12 × 72 mm, cotton-plugged, sterile
Syringes and needles, appropriate size
Alcohol, 70%, or tincture of iodine
Test tube racks, trays, or housing units

B.   METHODS

(1) *ANIMALS*
   (a) Prepare virus dilutions as described above.
   (b) With sterile technique, inoculate the test animal as described in
       Section B, the number per dilution depending on cost, etc. Do
       not forget to have control animals inoculated with the same
       volume of Hanks' medium or other diluent plus uninoculated
       controls.
   (c) Place in appropriate cages or isolation units. Observe daily.

(2) *TISSUE CULTURE*
   (a) Set up a rack with 10 small sterile test tubes. With sterile techni-
       que, add 0.9 ml (or other amount as required) of Hanks' medium
       to each tube.
   (b) Set up a tissue culture rack with 10 rows of tissue culture tubes,
       three tubes per row, and place six control tubes at the end (or
       use the same number of bottles or dishes of appropriate cells).

(c) Rapidly thaw the virus sample. Add 0.1 ml (or desired amount) to the first dilution tube. Discard the pipet. Refreeze the virus sample.

(d) Thoroughly mix the contents of tube No. 1 and remove 0.4 ml. Add 0.1 ml of this 10^{-1} virus dilution to each of the three culture tubes (or bottles, or dishes) of row 1. Dispense the last 0.1 ml into dilution tube No. 2, and discard the pipet.

(e) Mix the contents of tube No. 2, remove 0.4 ml, add 0.1 ml of this 10^{-2} dilution to each tissue culture tube (or bottle, or dish) in row 2, and dispense the last 0.1 ml into dilution tube No. 3. Discard the pipet.

In effect, what you are doing is saving on pipets by making your dilution in one set of tubes and inoculating that dilution into corresponding tissue cultures as you go along. This is called an *outside master dilution* (Fig. 4-10).

(f) Repeat the procedure above until you have inoculated all the tissue culture tubes (or bottles or dishes) in row 10. Label and incubate the cultures at 37°C.

(g) Observe for changes (CPE) each day for 7 days (or longer, depending on the virus used).

(3) *EMBRYONATING EGGS*

(a) Prepare virus dilutions as described above.

(b) With sterile technique, inject the proper amount of virus dilution into the egg (see Chapter 2). You may use a separate syringe for each dilution beginning at the most concentrated and pro-

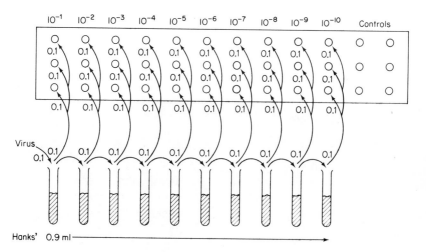

Fig. 4-10. Outside master dilution.

ceeding to the most dilute. We recommend the use of a single sterile syringe, and inoculation of the eggs beginning, however, at the most dilute, and working toward the most concentrated level of virus.

(c) Incubate eggs (in the proper position, depending on area inoculated) at 37°C.

(4) *OBSERVATION OF RESPONSES*

(a) *ANIMALS*   Observe animals daily. Responses will depend on the particular virus present. Note coughing, paralysis, inflammation of the eyes, etc. Check on the inoculated and uninoculated controls for similar symptoms.

(b) *CELL CULTURES*   Compare the virus-inoculated cultures with the uninoculated controls. Look for changes in the inoculates which are not visible in the controls, i.e., rounding of cells, thickening, tearing, etc.

(c) *EMBRYONATING EGGS*   Candle the eggs each day. Record any deaths and remove the eggs for harvest (Chapter 2). If embryo response is not death but an abnormality of some kind (paralysis, malformation), harvest the embryos at the appropriate time (Chapter 2) and record the abnormalities or changes observed in each embryo. Compare with control embryonating eggs harvested at the same time.

(5) *RECORDING RESPONSES*

(a) The "scoring" of each death or change is equivalent whether tubes, dishes, bottles, animals, or embryonating eggs are used.

(b) Score each tube, bottle, dish, embryo, or animal in each dilution as well as all controls. Record each as 0, $\pm$, 1+, 2+, 3+, 4+, for example:

Day	1	2	3
Dilution			
10^{-8}			
10^{-7}			
10^{-6}	000	000	000
10^{-5}	000	000	001
10^{-4}	000	000	111
10^{-3}	000	000	221
10^{-2}	001	112	223
10^{-1}	11$\pm$	222	334
Control	000	000	000

D. CALCULATION OF VIRUS TITERS

1. Introduction

The titer is usually expressed as the reciprocal of the highest dilution of the virus which causes a specified reaction in 50% of the material inoculated with, or exposed to, that dilution of infectious material. Where cell cultures are used as the indicator system, titers are expressed as the 50% tissue culture infective dose ($TCID_{50}$). Where experimental animals or eggs are used with death as the criterion, the lethal dose affecting 50% of the inoculates is expressed as the LD_{50}. It is obvious that a certain abbreviated vocabulary is in common usage and should be understood (Tables 4-1 and 4-2).

The method of Reed and Muench [39] is widely used for calculating the end point of viral activity (titer). The merits of this system are that, by accumulating deaths and survivals, reactors and non-reactors, over the entire dilution range, all the inoculates in the titration are used in the calculations rather than only those in dilutions near the end point. It is a relatively easy method to learn and has the advantage that it works. Once you have mastered Reed and Muench, you will, no doubt, wish to learn other methods of determining titers, such as that of Kärber [26] or that of Seligman and Mickey [50]. Whatever method you use, your percentage error will remain constant as long as

TABLE 4-1

Abbreviated Vocabulary Used in Expressing Virus Infectivity

Terminology	*Definition*
Indicator system	The susceptible animal, cell culture, plant, or other subject used to measure or indicate the activity of the virus.
I.U.	Infectious Unit. The smallest amount of virus which produces a recognizable reaction in the indicator system.
I.	Infective. Used where the measure of the response to viral exposure is a specified or preselected reaction, i.e., respiratory signs, skin reactions, paralysis, inflammation.
D.	Dose. Should be given as volume used as well as the dilution used in the preparation, i.e., 0.2 ml of a 1:100 (10^{-2}) dilution. The end point, or titer, of such material should be given so the number of infective units per unit volume is known.
Titer	Expressed as the reciprocal of that dilution of virus which affects at least 50%, but not more, of the indicator system exposed to that dilution. This is, therefore, the number of infectious units per unit volume.

TABLE 4-2

Examples of Terminology in Which Common Abbreviations are Found

Examples	Interpretation
"The $TCID_{50}$ is $10^{-6.7}/0.1$ ml"	This indicates that a virus suspension, diluted to $10^{-6.7}$ and inoculated in units of 0.1 ml, is the highest dilution which infected 50% of the tissue culture system used.
"The LD_{50} is $10^{-3}/0.2$ ml"	This indicates that a virus suspension, diluted to $10^{-3.0}$ and administered in units of 0.2 ml, is the highest dilution which infected 50% of the subjects so exposed.
"The virus titer is $10^{7.0}$"	The end point of viral activity is found in the 10^{-7} dilution. This dilution, then, is considered to have 1 (one) infectious unit.
"The virus titer is 10^6. To obtain 100 infectious units for inoculation, we use a dilution of 10^{-4}."	If the dilution of 10^{-6} has 1 (one) infectious unit (see example above), then the previous tenfold dilution (10^{-5}) would have 10 infectious units, and the tenfold dilution prior to that (10^{-4}) would have 10 times as much, or 100 units.

the same method is used consistently. You should always compare a titer with the titer of a control preparation assayed under the same conditions. Remember, the end point detects only virions able to initiate infection, and a titer refers to concentration of these infectious units.

2. Method

a. Record the response in the indicator system over a period of time which is suitable to both the virus under study and the indicator host. In the inoculation of cell cultures with some viruses, effects may not be apparent for several days and, as long as the cultures remain viable, you may find it necessary to record changes for several weeks. In the inoculation of mice with equine encephelomyelitis virus, for example, those which receive the more concentrated levels will be affected sooner than those inoculated with more dilute inoculum levels. You might, therefore, record the effects daily over a period of a number of days.

EXAMPLE

Five mice inoculated with X ml of virus "T" per dilution.
Observation period: 5 days.
Record your data in a form which resembles the following:

Readings on post—inoculation days:

Dilutions:		1	2	3	4	5
	10^{-5}	00000	00000	00000	00000	00000
	10^{-4}	00000	00000	00000	00000	00000
	10^{-3}	00000	00000	11000	33300	xxx00
	10^{-2}	00000	11000	33210	xxx3x	xxxxx
	10^{-1}	00000	11110	33330	xxxxx	xxxxx
Control		00000	00000	00000	00000	00000

Legend: Numbers indicate degree of paralysis.
X = death.
O = no effect, i.e., animal's appearance normal.

b. At the end of the test period, make note of the dilutions which span the range from every completely affected unit (animal, egg, cell culture) through to the dilutions which show no effect in any unit inoculated.

In the example above, the dilutions which span the 100 to 0% affected range by the end of the test period were 10^{-2} (100% affected), 10^{-3} (60% affected), and 10^{-4} (0% affected). The rate of infection is used as the criterion for determination of the titer of the virus in this example. Infectivity rates are expressed as a fraction in which the denominator represents the total number of mice, eggs, or cultures (tubes) inoculated and the numerator represents the number of mice, eggs, or cultures (tubes) affected by the virus. A reading of XXXXX in the 10^{-2} dilution is expressed as 5/5 meaning that five of the five mice inoculated were affected by the virus. As mentioned, the end point of the virus is indicated by the highest dilution which affected 50% or more of the inoculates. The titer of the undiluted sample of virus is written as the reciprocal of that highest dilution.

c. Determine the 50% end point as follows:

	Dilutions of Virus Spanning the 100% to 0% Affected Range (Example above)		
	10^{-2}	10^{-3}	10^{-4}
Infective (death) rate, each dilution	5/5	3/5	0/5
Number affected (deaths)	5	3	0
Number surviving (unaffected)	0	2	5
Accumulate (total) the *affected* mice (start at most dilute level; accumulate toward most concentrated level)	8 ←	= 5 + 3 ←	= 3 + 0

	Dilutions of Virus Spanning the 100% to 0% Affected Range (Example above)		
	10^{-2}	10^{-3}	10^{-4}
Accumulate (total) the *unaffected* mice (start at most concentrated level; accumulate toward most dilute level)	$0 \xrightarrow{\quad +2 = \quad} 2$		$\xrightarrow{\quad +5 = \quad} 7$
Express the mortality rate (No. *affected*/total)	8/8	3/5	0/7
Express per cent mortality	100	60	0

$$\uparrow$$
$$50\%$$

Obviously, the 50% level lies somewhere between the 10^{-3} and the 10^{-4} dilution.

      d. Compute the actual end point dilution (the dilution which would have given an exact 50% end point) using the following formula:

$$\frac{\text{per cent mortality next above } 50\% - 50\%}{\text{Per cent mortality next above } 50\% - \text{Per cent mortality next below } 50\%} = \text{proportionate distance}$$

Therefore, from the example above,

$$\frac{60\% - 50\%}{60\% - 0\%} = 0.2 \text{ (proportionate distance)}$$

Log of that dilution in which per cent mortality was next above 50% = -3.0. Then multiply the proportionate distance (0.2) × log of the dilution factor 10 ($= -1.0$) to obtain the real value of the proportionate distance $= -0.2$. (Since dilutions increase on a logarithmic scale, in a tenfold dilution, the log is 1.0; in a twofold dilution, the log is 0.3; in a fourfold dilution, the log is 0.6, etc.)

      Sum (50% end point):

$$-3.0 + -0.2 = -3.2$$

      e. You can then express the titer as

$\text{I.D.}_{50}/0.1$ ml is $10^{-3.2}$ or 1.58 (which is the antilog of 0.2) × 10^3

f. Therefore, 10 I.D.$_{50}$ is $10^{-2.2}$, or 1.58×10^2, and 100 I.D.$_{50}$/0.1 ml is $10^{-1.2}$, or 1.58×10^1.

g. The following example, like that discussed above, is a hypothetical one. This time, cell cultures represent the host but the method is equally valid with any indicator system (Table 4-3). In this example, readings were recorded on the second, fourth, and sixth days postinoculation, and the progressive effect of the virus on the cell sheets was noted. On the second day, no effect was observed beyond the 10^{-4} dilution, in which all four tubes were recorded as "S." This indicates a slight alteration but is not definite enough to be regarded as a positive response in making calculations. Readings of 1 or greater indicate a degree of positive cellular response which eventually progressed to stage 4 or total destruction. Increasing stages of cell degeneration were observed in the cultures inoculated with the $10^{-3}, 10^{-2},$ and 10^{-1} dilutions of the virus. On the fourth day, the effect of the virus was seen in the tubes inoculated with the 10^{-6} dilution but not in those inoculated with the 10^{-7} dilution or beyond. On the sixth day, the effect was essentially the same, except that in the cultures inoculated with the more concentrated dilutions of the virus, progressing lethal effects were observed.

The per cent of the cultures affected and the approximation of the TCID$_{50}$ are calculated from this example as follows:

Virus Dilution:	10^0	10^{-1}	10^{-2}	10^{-3}	10^{-4}	10^{-5}	10^{-6}	10^{-7}	10^{-8}
Number affected:	$\frac{4}{4}$	$\frac{4}{4}$	$\frac{4}{4}$	$\frac{4}{4}$	$\frac{4}{4}$	$\frac{4}{4}$	$\frac{3}{4}$	$\frac{0}{4}$	$\frac{0}{4}$
TCID$_{50}$:	10^6	10^5	10^4	10^3	10^2	10^1	10^0		
Mortality rate:							100%	75%	0%

It is apparent that the 50% end point is found in a dilution higher than 10^{-6} but less than 10^{-7}. For practical purposes, it is sufficient to state that the titer of this sample is indicated by the 10^{-6} dilution of the virus (1 : 1,000,000) and that the undiluted virus sample contains approximately 10^6 (1,000,000) TCID$_{50}$ per 0.1 ml of inoculum. It is evident that a 10^{-4} dilution (1 : 10,000) contains 10^2 (100) TCID$_{50}$ per 0.1 ml, and a 10^{-1} dilution (1 : 10) contains 10^5 (100,000) TCID$_{50}$ per 0.1 ml.

A more accurate assessment of the results obtained is made by taking into consideration the number of accumulated affected and noninfected cultures. Again, using the data from this hypothetical example, the results may be analyzed in tabular form as given in the previous example (2c).

The 50% end point is found in a dilution which is greater than 10^{-6} (where three-quarters, or 75%, of the cultures were affected) but less than 10^{-7} (where none, or 0%, of the cultures were affected). The proportionate distance at which the 50% end point occurs between these two dilutions would be

$$\frac{75 - 50}{75 - 0} = \frac{25}{75} = 0.33 = 0.3 \text{ (rounding to nearest tenth)}$$

TABLE 4-3

Procedure for Hypothetical Virus Titrations

Tube number	Amount diluent (ml)	ml/tube virus	Procedure for mixing	Resultant dilution	Amount ml inoculated into each of 4 tubes	Readings on postinoculation days			Ratio affected
						2	4	6	
1	9.0	1.0	Mix, transfer 0.1 ml to next tube; discard pipet	10^{-1}	0.1	2233	4444	XXXX	4/4
2	9.0	None	Repeat, as above	10^{-2}	0.1	2221	4444	XXXX	4/4
3	9.0	None	Repeat, as above	10^{-3}	0.1	1221	4342	4444	4/4
4	9.0	None	Repeat, as above	10^{-4}	0.1	SSSS	2232	2444	4/4
5	9.0	None	Repeat, as above	10^{-5}	0.1	0000	2111	2232	4/4
6	9.0	None	Repeat, as above	10^{-6}	0.1	0000	0211	S322	3/4
7	9.0	None	Repeat, as above	10^{-7}	0.1	0000	0000	0S00	0/4
8	9.0	None	Repeat, as above	10^{-8}	0.1	0000	0000	0000	0/4

Symbols used in reading the inoculated cultures:

0 = cell sheet appears normal.

S = suspicious, no definite CPE, but sheet seems slightly altered; not considered a positive identification of virus activity.

1 = first signs of CPE, about 25% of the sheet affected.

2 = 50% of the sheet affected.

3 = 75% of the sheet affected.

4 = 100% of the sheet affected.

X = a non-readable tube culture—sheet missing.

When tenfold dilutions are used, the $TCID_{50}$ (titer of the virus) is expressed as the reciprocal of the highest virus dilution affecting 50% or more of the cultures inoculated with that dilution (in this example, 10^6) plus the factor (0.3) which indicates the amount this dilution would have to be increased to affect exactly 50% of the cultures. The $TCID_{50}$ of the virus would then be written as $10^{6.3}$ per unit volume inoculated.

Serologic Techniques

FIVE

A. NEUTRALIZATION TESTS

1. Introduction

In a neutralization test, serum and virus are brought together under certain conditions and are inoculated into a susceptible host (animal, egg, tissue culture). If antibodies for the virus in question are absent, disease, lesions, or death may result. When antibodies are present, no such reactions are noted.

The virus sample may consist of the various fluids from embryonated eggs, tissue culture fluids, or extracts of infected tissue (brain, liver, etc.). These may be used directly or after low speed centrifugation (205g, 15 to 20 min) to remove debris. These preparations should be used fresh when possible, or the tissues may be frozen: if lyophilized, at $-20°C$; if frozen in ampuls, at $-60°C$ or lower.

The selection of the indicator system will depend on the infectiousness and lethality of the virus for a particular host, the cost, the availability, the ease of handling, etc. Animals (mice, hamsters, chicks), embryonated eggs (hen, duck), or cells in culture may be employed.

There are two methods of performing neutralization tests. In the *alpha* procedure, equal quantities of a constant amount of serum and increasing dilutions of virus are incubated together and then inoculated into the indicator system. In the *beta* procedure, virus at a concentration of 10 and/or 100 $TCID_{50}$ is incubated with twofold serial dilutions of serum before inoculation.

94

The latter method is more generally used with tissue culture because it is more economical in the use of sera and gives a greater delicacy of test. Also, a broàder range of antibody titer can be measured where a constant amount of virus is used. The temperature and length of incubation of the serum-virus mixture will depend on the agents used, although there appears to be little agreement as to the best temperature to use in each particular circumstance or even the need for incubating the serum-virus mixture.

2. Alpha Procedure

Although cell cultures are used as the indicator system in the procedures described below, it should be understood that the same procedures apply equally to doing these tests in animals or eggs. One simply substitutes the appropriate host for the cell culture tubes described below.

a. MATERIALS

> Frozen virus suspension
> Virus-specific antiserum
> Uninoculated tissue cultures: three tubes per dilution for
>   virus-serum test and virus titer control, plus six uninoculated
>   controls
> Hanks' medium
> Test tubes, 12 × 72 mm, sterile
> Pipets, 1.0 and 10.0 ml, sterile
> Tissue culture racks
> Test tube racks
> Water baths, 37 and 56°C
> Bunsen burner

b. METHOD

(1) Inactivate serum at 56°C for 30 min to destroy heat-labile virus inhibitors. Dilute the serum 1:10 in Hanks' medium.

(2) Set up 10 small sterile test tubes in a rack. Add 0.5 ml of 1:10 serum to each tube.

(3) Set up a tissue culture rack with 10 rows of tissue culture tubes, 3 tubes per row, and place 6 control tubes at the end.

(4) Set up another set of 10 small sterile test tubes. Using the outside-master technique, prepare ten-fold serial dilutions of virus in Hanks' medium, adding 0.5 ml of each virus dilution to each corresponding tube of 1:10 serum and 0.1 ml to each replicate tissue culture tube for a virus titration (Fig. 5-1).

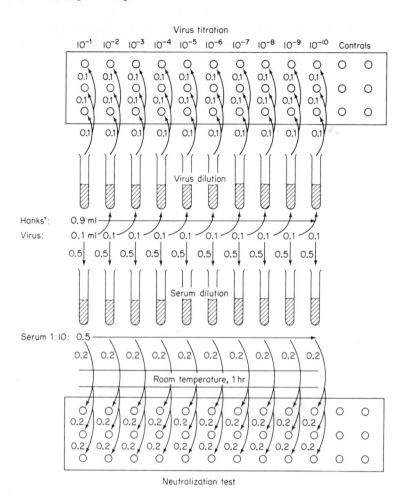

Fig. 5-1. Neutralization test.

(5) Label rack containing control virus titration and place in 37°C incubator.

(6) Shake all serum-virus tubes. Incubate 1 hr at room temperature.

(7) In the meantime, set up another tissue culture rack, 10 rows of tissue culture tubes, 3 tubes per row, for the neutralization test.

(8) At the end of the incubation period, inoculate 0.2 ml of serum-virus mixture into the replicate tissue culture tubes set up for this purpose, using a different pipet for each serum-virus dilution (Fig. 5-1). Label rack and place in 37°C incubator.

Note that in the virus titration tubes you inoculated 0.1 ml per tube. Now, when you are inoculating the material for the neutralization test, you are using 0.2 ml per tube. If it is remembered that the inoculum for the neutralization test is half virus and half serum, an adjustment (0.1 ml of virus and 0.1 ml serum equals 0.2 ml total) must be made in inoculum volume to assure that each unit of the indicator system receives an amount of virus which matches that of the titration containing no serum.

(9) Observe for CPE on the fourth through the seventh day and record results as previously described.

3. Neutralization Index

In the *alpha* neutralization test (constant serum plus virus dilutions), the neutralizing effect of the serum is expressed by the *neutralization index*. This is determined by titering the virus in the presence of diluent and in the presence of the test serum, calculating the LD_{50}, and then subtracting the exponent of the latter from the former, disregarding the negative sign. The number obtained represents the logarithm of the neutralization index.

EXAMPLE

LD_{50} of control titer $= 10^{-6.5}$
LD_{50} of neutralization titer $= 10^{-4.5}$
Neutralization index $= 10^{(6.5-4.5)} = 10^{2.0} = 100$
Log of neutralization index $= \log 100 = 2.0$

In general, titrations done in ten-fold dilutions have been found to vary by a factor of 10, or $\log_{10} = 1$. Therefore, neutralization indices of less than 10 (log less than 1) are considered not significant; values between 10 and 50 (logs 1.0 to 1.6) are questionable; indices over 50 (log 1.7 or greater) are significant.

4. Preparation of Exact Log Dilution

By this time, you should be well aware of the titer of the virus with which you have been working. In order to perform the *beta* neutralization test, it is necessary to prepare 10 and/or 100 $TCID_{50}$ of virus in exact log dilutions.

EXAMPLE

Let's say that your virus normally titers at $10^{-6.26}$.
This means that $10^{-5.26}$ is 10 $TCID_{50}$, and $10^{-4.26}$ is 100 $TCID_{50}$.
You want to run a neutralization test with 10 $TCID_{50}$ of virus, i.e., a $10^{-5.26}$ dilution.
To prepare exactly $10^{-5.26}$ of virus,
  a. Make a regular virus dilution to $10^{-5.0}$.
  b. Find the antilog of 0.26 (which is 1.82).
  c. $1.82 - 1.00$ (virus) $= 0.82$; therefore, take 1 part of 10^{-5} dilution of virus plus 0.82 part of diluent.

5. Beta Procedure

a. Inactivate serum at 56°C, 30 min.

b. Set up 5 sterile test tubes containing 0.5 ml Hanks' medium. Add 0.5 ml serum to tube No. 1 and make twofold serial dilutions from there.

c. Thaw virus rapidly. Prepare 10 and 100 $TCID_{50}$ of virus in exact log dilutions.

d. Add 0.5 ml of 10 or 100 $TCID_{50}$ of virus to each serum dilution tube with sterile technique. Mix. Incubate at room temperature for 1 hr.

e. Inoculate 0.2 ml of serum-virus mixture into replicate tissue cultures.

f. Carry a standard virus titer plus uninoculated tissue culture controls (Fig. 5-2).

g. Observe for CPE on the fourth through the seventh day.

h. Determine the neutralization index.

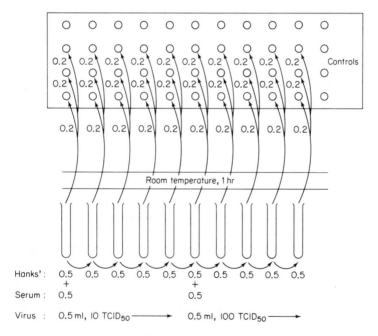

Fig. 5-2. Beta procedure.

B.   HEMAGGLUTINATION

Some viruses, or an antigen derived from them, adsorb to red blood cells through receptor sites on their surface. As a result, the erythrocytes agglutinate, a phenomenon termed *hemagglutination* (Table 5-1).

The hemagglutination (HA) test is a simple one. Basically, red cells in suspension are brought in contact with (1) a preparation containing virus (or its antigen) or (2) a preparation without virus or its antigen. The red cells sediment out of solution by gravity (1) in the presence of virus (or its antigen) in large clumps or (2) in the absence of virus as a packed disc in the bottom of the test tube. This is called the *pattern method* and was devised by J. SALK [45].

Another more technically exacting method based on sedimentation will not be considered here [12]. That is not to say that there are no problems in the execution of the pattern method, however. There are limitations of pH, temperature, range of agglutinable red cells, non-specific serum inhibitors, etc.

The species of animal whose red cells will be agglutinated will vary depending on the virus. Some viruses have a rather broad range, others have a very narrow range of species whose red cells they agglutinate. With enteroviruses, human O erythrocytes are used; with Newcastle, influenza A and B, and para-influenza 1-4, human O, guinea pig, and chicken red blood cells are used; with reoviruses, human O or bovine cells; with adenoviruses group 1, rhesus monkey cells; and with adeno groups 2 and 3, rat red blood cells, for example.

A concentration of packed red cells ranging between 0.25 and 1.0% (final concentration in test of 0.12 to 0.33%) must be used. There is an inverse relationship between erythrocyte concentration and agglutination titer of a virus antigen. Mammalian cells are normally used at a slightly higher concentration than avian cells because they sediment more slowly; increasing the number decreases the sedimentation rate.

Red cells are often purchased from commercial sources. If you must bleed an animal, however, use 1 part of 5% sodium citrate to five parts of blood. Mix the blood-citrate solution by repeated inversions of the stoppered tube or flask used to collect the blood. Then wash the cells three times in 0.85% saline: gently mix the citrate-blood suspension with approximately five volumes of the saline, sediment the suspension at 205 g for 10 min, pour off the supernate, add fresh saline, and repeat twice more. Drain off the last wash. Resuspend the washed cells in saline to give a 10% stock solution. This may be kept at refrigerator temperature (4°C) for 5 or 6 days.

If you wish to keep cells for 2 weeks, it is suggested that the animal be bled by syringe containing 1.5 ml of acid-citrate-dextrose (ACD) for each 8.5 ml of blood. Wash with dextrose-gelatin-veronal (DGV), first with 1 volume of blood to 2.5 volumes of DGV and then 1 volume of blood to 3 volumes of DGV, at 205 g, in the cold (4°C). Resuspend in 3 volumes DGV (roughly 8% to 10% by volume RBC (red blood cells) or a 1:40 dilution, or that dilution which gives an optical density of 0.450 at 490 mμ). Keep this stock suspension at 4°C.

With most viruses, a pH from 6.0 to 8.5 is satisfactory, and the test is made in either 0.85% NaCl or 0.01 M PBS, pH 7.2.

TABLE 5-1

Hemagglutination by Certain Animal Viruses

Virus	Source	RBC	%	Temp. (°C)	Time (hr)	HAd
Enterovirus	Tissue culture fluid	Human 0	0.5	4, 22, and/or 37	1–2	—
Coxsackie A						
B-3	"	"	"	4		
B-1-3-5	"	"	"	20		
Arbovirus	Brain-infected mouse tissue	Newborn 0 1 day old chick or goose pH 6–7	"	37 4, 20, 37, or room temperature	—	
Measles	Tissue culture fluid	Monkey (*Cercopithicus*)	1.0	37	1 If +, continue 4 hr or longer for hemolysis	+
Reovirus						
Type 1	Tissue culture fluid	Human 0	0.5	20	1–2	—
2	"	"	"	"	"	
3	"	Bovine	"	4	"	
Myxovirus						
Influenza A, A₁, A₂	Tissue culture fluid, amniotic-allantoic fluid pools	Chick G.P. (Guinea pig) Human 0	" " "	4, 22, 37 4, 20 20	" " "	+
B	"	Chick G.P. Human 0	" " "	4, 22 4, 20 20	" " "	
C	"	Chick Human 0	" "	4 4	" "	
Para-influenza Type 1, 3	"	Chick	"	4, 22	"	

100

TABLE 5-1 (Continued)

Virus	Source	RBC	%	Temp. (°C)	Time (hr)	HAd
		G.P.	"	4, 20	"	
		Human 0	"	4, 20	"	
2	"	Chick	"	4	"	
		G.P.	"	4, 20	"	
		Human 0	"	20	"	
4	"	Rhesus monkey	"	20, 37	"	
		G.P.	"	20, 37	"	
5	"	Chick	"	4, 22	"	
		G.P.	"	4, 20	"	
		Human O	"	20	"	
Adenovirus						
I	Tissue culture fluid	Rhesus monkey	"	20	"	—
II	"	Rat	"	37	"	
II-A	"	Rat	"	37	"	
III	"	Rat ±	"	37	"	
IV	"	—				

Temperatures of the hemagglutination reaction are usually 22 or 37°C because sedimentation is faster and patterns are sharper. Some viruses agglutinate only at 4°C (influenza C, for example).

The time required for sedimentation of erythrocytes will depend on the species. Avian cells at 37°C will settle in 1 hr, whereas mammalian cells may take anywhere from 4 to 8 hr.

1. Materials

> Virus suspension
> Citrated red blood cells
> PBS, 0.01 M, pH 7.2
> Centrifuge tubes
> Test tubes, 12 × 72 mm, round-bottomed
> Pipets, 1.0 and 10.0 ml, sterile
> Centrifuge
> Water bath, 37°C
> Rack

2. Methods

   a. Separate plasma from red blood cells by centrifuging citrated cells as described above.

   b. Prepare a 0.5% suspension of red blood cells in PBS.

   c. Prepare twofold serial dilutions (1 : 10 to 1 : 640) of virus suspension in PBS (Fig. 5-3).

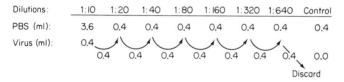

Fig. 5-3. Preparation of virus dilutions.

   d. Add 0.4 ml of 0.5% RBC to all tubes. (Control tubes should contain 0.4 ml of PBS plus 0.4 ml of RBC).

   e. Shake tubes. Place in a 37°C water bath or at the appropriate temperature for the virus concerned until control cells have completely settled.

3. Test Results

Test results are determined by the distinct pattern formed by the sedimented cells. Patterns are read by viewing the tubes from the bottom or by holding the rack over a mirror.

a. In the control tube, cells roll to the bottom and settle as a sharply outlined disc.
b. Intermediate reactions are seen as irregular clumps of cells associated with a halo of finely aggregated cells.
c. Maximal agglutination is characterized by a uniform salmon-pink film covering the bottom of the tube.

EXAMPLE

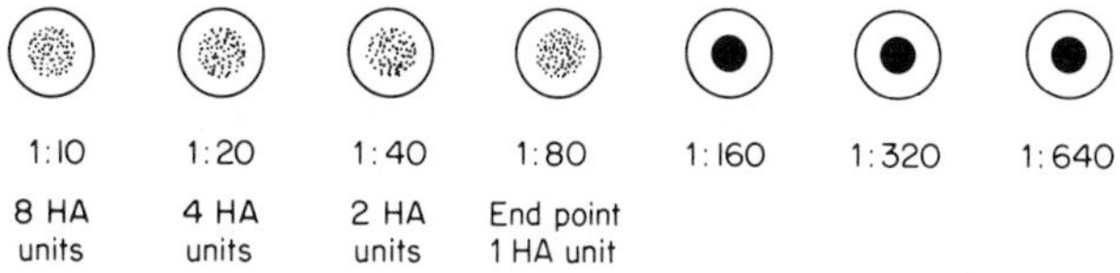

1:10	1:20	1:40	1:80	1:160	1:320	1:640
8 HA units	4 HA units	2 HA units	End point 1 HA unit			

Fig. 5-4. Illustration of HA bottom patterns.

d. Hemagglutination results are normally graded 0, 1, 2, 3, or 4+.
e. The end point is taken to be the highest dilution of virus suspension producing maximal agglutination (4+). This dilution is said to contain 1 HA (hemagglutinating) unit per whatever volume you used. The next lower dilution (higher concentration) is said to contain 2 HA units, etc.

C. HEMAGGLUTINATION—INHIBITION

The hemagglutination inhibition (HI) test is based on the inhibition of viral hemagglutinin by specific antibody. The test is used to detect, indirectly, the presence of a hemagglutinating virus. Because the virus is a foreign protein, it will elicit the formation of antibodies in the host. If serum of the host inhibits hemagglutination by a virus which normally does so, then this means that the virus was in the host.

1. Materials

Virus suspension
Specific antiserum
Citrated red blood cells
PBS, sterile
Centrifuge tubes, sterile
Test tube rack
Test tubes, 12 × 72 mm, round-bottomed, sterile
Pipets, 1.0 and 10.0 ml, sterile
Centrifuge
Water bath, 37°C

2. Method

a. Make twofold serial dilutions of antiserum in PBS, beginning at 1: 10. Distribute the serum dilution in 0.2-ml amounts in small test tubes.

b. Titrate the antigen as in the HA test. Determine the dilution which contains 4 units per 0.2 ml.

EXAMPLE

If the HA titer of the virus is 1: 80, this represents 1 HA unit.

Thus, a dilution of 1: 10 would contain 8 HA units in 0.4 ml or 4 HA units in 0.2 ml.

Add 4 units of antigen in 0.2 ml volume to each antiserum tube.

c. Include the following controls:
   (1) Serum alone, 1: 10 dilution, 0.4 ml.
   (2) Antigen alone, 4 units, 0.4 ml.
   (3) Known positive serum, 1: 10, 0.4 ml.
   (4) Known negative serum, 1: 10, 0.4 ml.
   (5) As a check on dilution error or variations in red blood cells from one batch to another, run a back-titration on the 4 HA units per 0.2 ml dilution:
      (a) Add 0.4 ml PBS to each of five test tubes.
      (b) To the first tube, add 0.4 ml of the virus dilution (in the case above, 1: 10) containing 8 HA units. Mix and transfer 0.4 ml to the second tube. Continue and discard 0.4 ml from the last tube.

d. Incubate at room temperature for 30 to 60 min.

e. Add 0.4 ml of 0.5% red blood cells to all tubes. Shake. Place all tubes at the appropriate temperature until control cells have completely settled.

f. Read the serum titer. This is the highest dilution of serum inhibiting hemagglutination by the virus and is expressed as the reciprocal of that serum dilution.

EXAMPLE

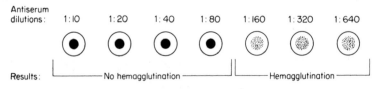

Fig. 5-5. Illustration of HI serum titer.

g. Read the back-titration as follows:
   Back-titration:

Fig. 5-6. Illustration of back titration.

Only the first three tubes should show agglutination indicating that there were actually 4 HA units per 0.2 ml used in the test. If more than three tubes show agglutination, the virus concentration must be changed (diluted) by the addition of PBS. If less than three tubes show agglutination, the virus concentration must be increased by the addition of more virus.

h. Read the controls:

Fig. 5-7. Illustration of controls.

(1) Serum alone should be negative.
(2) Antigen alone should be negative.
(3) Known positive serum should be positive.
(4) Known negative serum should be negative.

D. HEMADSORPTION TEST

Specific erythrocyte adsorption to tissue culture cells provides a practical approach to diagnosis of virus infection before CPE is observed or in cases where agents may propagate without apparent damage to such cells. The technique described was developed by VOGEL AND SHELOKOV [51, 56]. The test may be done with human O, calf, guinea pig, sheep, rat, rhesus monkey, and chicken red blood cells at 37°C, room temperature, or 4°C.

1. Materials

Infected tissue culture cells
Control tissue culture cells
Appropriate red blood cells, citrated
PBS, sterile
Pipets, 1.0 and 10.0 ml, sterile

Centrifuge tubes, sterile
Flask, sterile

2. Method

a. Wash the citrated red blood cells three times with PBS. Prepare a 0.4% suspension of washed cells in PBS.
b. Pour the media from control and infected (before CPE is visible) tissue cultures into a contamination pan.
c. Add 0.2 ml of the 0.4% suspension of red blood cells to each tube. Place the tubes in a horizontal position such that the blood cells cover the surface of the cell sheet and incubate at appropriate temperature.
d. Observe microscopically at intervals for clumping of the red blood cells and visible adsorption of the blood cells to the cell sheet.

E. UNMASKING OF HEMAGGLUTININS BY FLUOROCARBON

Tissue culture preparations may vary with respect to hemagglutinating activity, and it has been found that inhibitors may be present in infected fluids which mask the HA reaction. SCHMIDT *et al.* [48] found that better HA antigens could be produced by infecting cell cultures with large virus inocula followed by harvesting them 24 hr later. Treatment of such harvests with the fluorocarbon Genetron 113 "unmasks" any hidden hemagglutinins.

1. Materials

24-hr infected tissue cultures
Genesolv-D (trifluorotrichlorethane, $CCl_2F–CClF_2$)
Servall homogenizer, cold, sterile
PBS, sterile
Red blood cells (appropriate type)
Pipets, 1.0 ml and 10.0 ml, sterile
Test tubes, 12 × 72 mm, sterile
Centrifuge tubes, sterile
Test tube rack
Bunsen burner

2. Method

a. FLUOROCARBON TREATMENT

(1) Twenty-four hours post virus inoculation, freeze and thaw the cultures three times. Harvest the fluids.

(2) To the harvested fluids, add an equal volume of Genesolv-D. Homogenize at 44,000 RPM for 3 min, in the cold. Centrifuge at 205 g for 10 min.

(3) Remove the supernatant fluid to a sterile test tube. Label with the contents, treatment, and data.

b. TEST

(1) Prepare twofold serial dilutions of the supernatant fluid in PBS. Add 0.4 ml of each dilution to triplicate small sterile test tubes.

(2) Wash appropriate red blood cells three times with sterile PBS. Make a 1 % suspension in PBS and add 0.1 ml to each tube of the test.

(3) Shake all tubes. Place one set at 4°C, one set at room temperature, and one set at 37°C for 1 hr or until all the red cells have settled.

(4) Read the bottom patterns as indicated previously.

F.   THE COMPLEMENT FIXATION TEST

1. General Information

a.   THE ANTIGEN-ANTIBODY-COMPLEMENT SYSTEM

(1) Virus suspension prepared from undiluted, stored materials (infected cell cultures, egg fluids, or ground animal tissues) acts as antigen. Diluted with veronal buffer solution (pH 7.3–7.4), it is used as 2 antigenic units based on titration with standard positive serum.

(2) Serum should be obtained from blood samples which have been drawn under aseptic conditions, and every precaution should be taken to avoid hemolysis. Inactivated serum (56°C, 30 min), diluted 1 : 5 in veronal buffer, is tested for the presence of specific antibody.

(3) Normal guinea pig serum (commercially prepared) is used as complement. It is diluted in buffer so that each 0.5 ml contains 2 exact units.

b.   THE HEMOLYTIC SYSTEM

The hemolytic system consists of complement plus 1.0 ml of equal parts of 2% washed sheep red blood cells and hemolytic amboceptor (antisheep hemolysin) containing 2 units per 0.5 ml. The mixture is held at 37°C for 10 min before use in the test.

2. Materials

Sheep red blood cells
Amboceptor (antisheep hemolysin)

Complement
Antigen
Antiserum
Test tubes, 12 × 72 mm
Centrifuge tubes
Pipets, 0.1, 1.0, 5.0, and 10.0 ml
Test tube rack
Centrifuge

a. SHEEP RED BLOOD CELLS

(1) Mix the commercially prepared sheep red blood cells by gentle rotation of the flask. Aseptically remove sufficient blood by means of a sterile pipet and place in a graduated conical centrifuge tube.

(2) Centrifuge for 10 min at 1000 RPM (205 g). Decant the supernatant fluid by suction through a capillary pipet, taking off the upper, white layer (buffy coat) of cells.

(3) Refill the tube with veronal buffer, resuspend cells by gently inverting the tube. Repeat the centrifugation and washing twice more. If the supernatant fluid is not colorless after the third washing, cells are too fragile and should not be used. Fresh sheep blood should be obtained.

(4) Suspend the packed cells in veronal buffer. Keep the cells cold by placing in an ice bath.

b. COMPLEMENT

(1) Prepare a 1:25 dilution of complement by adding 0.5 ml of rehydrated complement to exactly 12.0 ml of cold veronal buffer. Keep the working solution cold in an ice bath.

(2) Follow the protocol shown in Table 5-2. Place the rack of tubes in a water bath at 37°C for 45 min.

(3) Read the test immediately at the end of the incubation period. The *exact unit* is the smallest amount of diluted complement (1:25) showing complete and sparkling hemolysis in 45 min at 37°C. Complement is then diluted with veronal buffer so that 2 exact units are contained in a volume of 0.5 ml. The following equation is used to determine the amount of veronal buffer necessary to add to complement so that each 0.5 ml contains 2 exact units:

$$\frac{\text{The dilution of complement}}{\substack{\text{Number of units of complement} \\ X \text{ the volume in which those} \\ \text{units occur}}} = \frac{\text{``}X\text{''} \text{ (i.e., the final dilution to prepare)}}{\substack{\text{Number of milliliters of complement} \\ \text{containing 2 exact units}}}$$

TABLE 5-2

Complement Titration

Reagents	Tube numbers									
	1	*2*	*3*	*4*	*5*	*6*	*7*	*8*	*9*	*10*
Buffer (ml)	1.45	1.425	1.4	1.375	1.35	1.325	1.3	1.275	1.250	1.5
Complement (1:25 or 4%) (ml)	0.05	0.075	0.1	0.125	0.15	0.175	0.2	0.225	0.250	—
Amboceptor, 2 units per 0.5 ml	0.5	0.5	0.5	0.5	0.5	0.5	0.5	0.5	0.5	0.5
Sheep red blood cells (2%) (ml)	0.5	0.5	0.5	0.5	0.5	0.5	0.5	0.5	0.5	0.5

EXAMPLE

If the unit in Table 5-2 is read at tube No. 5 (0.15 ml of 1 : 25 complement), then

$$\frac{25}{2 \times 0.15} = \frac{X}{0.5}$$

$$X = 41.6$$

Thus, 1.0 ml of complement plus 40.6 ml of veronal buffer would contain 2 exact units of complement per 0.5 ml.

c. AMBOCEPTOR (ANTISHEEP HEMOLYSIN) TITRATION

  (1) Prepare a 1 : 1000 dilution of amboceptor as follows:
      (a) Add 0.1 ml antisheep hemolysin (commercially prepared) to 9.9 ml of veronal buffer (1 : 100 dilution).
      (b) Add 1.0 ml of 1 : 100 dilution to 9.0 ml of veronal buffer (1 : 1000 dilution).
  (2) Follow the protocol shown in Table 5-3 and prepare amboceptor dilutions as shown.
  (3) Transfer 0.5 ml of each dilution, starting with the preliminary dilution of 1 : 1000, to each of 11 tubes (Table 5-4).
  (4) Add complement (2 units per 0.5 ml), 1.0 ml of buffer, and 0.5 ml of 2% sheep cells to each tube. Place the rack of tubes in a 37°C water bath, incubate for 45 min, and read immediately. The exact unit of amboceptor is defined as the highest dilution (smallest

TABLE 5-3

Hemolysin Dilution for Titration

Tube number	Dilutions	Plus	Veronal buffer	Equals	Final dilution
1	1.0 ml of 1 : 1000		1.0 ml		1 : 2000
2	1.0 ml of 1 : 1000		2.0 ml		1 : 3000
3	1.0 ml of 1 : 1000		3.0 ml		1 : 4000
4	1.0 ml of 1 : 1000		4.0 ml		1 : 5000
5	1.0 ml of 1 : 3000		1.0 ml		1 : 6000
6	1.0 ml of 1 : 4000		1.0 ml		1 : 8000
7	1.0 ml of 1 : 5000		1.0 ml		1 : 10,000
8	1.0 ml of 1 : 6000		1.0 ml		1 : 12,000
9	1.0 ml of 1 : 8000		1.0 ml		1 : 16,000
10	1.0 ml of 1 : 10,000		1.0 ml		1 : 20,000

TABLE 5-4

Titration of Hemolysin

Tube number	Hemolysin (0.5 ml of each)	Complement (2 exact units)	Buffer (ml)	Sheep cells (2%) (ml)
1	1:1000	0.5 ml	1.0	0.5
2	1:2000	0.5 ml	1.0	0.5
3	1:3000	0.5 ml	1.0	0.5
4	1:4000	0.5 ml	1.0	0.5
5	1:5000	0.5 ml	1.0	0.5
6	1:6000	0.5 ml	1.0	0.5
7	1:8000	0.5 ml	1.0	0.5
8	1:10,000	0.5 ml	1.0	0.5
9	1:12,000	0.5 ml	1.0	0.5
10	1:16,000	0.5 ml	1.0	0.5
11	1:20,000	0.5 ml	1.0	0.5
12	None	0.5 ml	1.5	0.5

amount) causing complete and sparkling hemolysis of 0.5 ml of a 2% suspension of sheep red blood cells in 45 minutes at 37°C in the presence of 2 exact units of complement.

EXAMPLE

If the last tube showing complete and sparkling hemolysis at the end of the incubation period is tube No. 6 (1:6000 dilution, Table 5-4), this is the exact unit. Two exact units would then be tube No. 3, or a 1:3000 dilution of hemolysin (0.5 ml contains 2 exact units).

d. ANTIGEN TITRATION

(1) *Reagents Required*
   (a) Antigen diluted in veronal buffer to the approximate range of activity.
   (b) Complement, 2 exact units in 0.5 ml.
   (c) Amboceptor, 2 exact units in 0.5 ml.
   (d) Sheep red blood cells, 2% suspension.
   (e) Standard positive and negative sera, heated at 56°C, 30 min.

(2) *Method*
   (a) Set up the protocol as shown in Table 5-5.
   (b) Shake the rack of tubes to ensure thorough mixing and place in a 37°C water bath for 1 hr, or use the long fixation test (overnight at 4 to 6°C; then 37°C, 10 min).

TABLE 5-5

Antigen Titration

Reagents										Rear row of tubes—tube number									
	1	2	3	4	5	6	7	8	9	10	11	12	13	14	15	16	17	18	19
Buffer (ml)	0.875	0.85	0.8	0.75	0.7	0.65	0.5	0.4	0.9	0.85	0.8	0.7	0.6	0.5	0.4	0.6	0.9	1.0	1.5
Antigen (ml)	0.025	0.05	0.1	0.15	0.2	0.25	0.4	0.5	—	0.05	0.1	0.2	0.3	0.4	0.5	0.5	—	—	—
Positive serum (ml)	0.1	0.1	0.1	0.1	0.1	0.1	0.1	0.1	0.1	0.1	0.1	0.1	0.1	0.1	0.1	—	0.1	—	—
Complement (ml)	0.5	0.5	0.5	0.5	0.5	0.5	0.5	0.5	0.5	0.5	0.5	0.5	0.5	0.5	0.5	0.5	0.5	0.5	—
								Incubate for 1 hr in 37°C water bath											
Sensitized sheep RBC (ml)	1.0	1.0	1.0	1.0	1.0	1.0	1.0	1.0	1.0	1.0	1.0	1.0	1.0	1.0	1.0	1.0	1.0	1.0	1.0
								Incubate for 45 min in 37°C water bath											

(c) During this incubation period, mix the 2% sheep red blood cell suspension with an equal quantity of amboceptor containing 2 units per 0.5 ml. Sensitize this mixture in a 37°C water bath for 10 min prior to use.

(d) Add 1.0 ml of the sensitized red blood cells to each tube; shake the rack of tubes and return it to the water bath for 45 min.

(e) Read the results immediately at the end of the incubation period. To facilitate reading of the test, the tubes may be centrifuged for 5 min at 205 g and then read, or the entire rack may be placed immediately into a refrigerator and the test read the next morning.

The antigenic unit is defined as the smallest amount of diluted antigen which will completely fix (4+ reaction) 2 units of complement in the presence of standard positive serum. The anticomplementary unit is the amount of diluted antigen which inhibits hemolysis in the presence of standard negative serum.

(3) *Interpretation of Titration*

Tube number	Controls	Reaction
9	Positive serum	Hemolysis
16	Antigen	Variable hemolysis
17	Negative serum	Hemolysis
18	Complement	Hemolysis
19	Amboceptor plus RBC	No hemolysis

In tubes No. 1 through 8 you should observe a spectrum of hemolytic activity from complete hemolysis to complete fixation, i.e., no hemolysis [complete fixation should fall somewhere between tubes No. 2 and 5 (0.05 to 0.2 ml)]. If complete fixation is observed in every tube containing positive serum except in the positive serum control (tube No. 9), evidently the antigen unit is stronger than the chosen dilution of antigen, and a higher dilution must be tried until the range from complete hemolysis to fixation (4+) is obtained in the test. Thus, if the first tube showing complete fixation (4+) is No. 3, the unit is 0.1 ml of the diluted antigen. To calculate the dilution of antigen to contain 2 exact units per 0.5 ml, use the following proportion:

$$\frac{\text{Dilution of antigen}}{2 \text{ units}} = \frac{X}{\text{volume of test done (0.5 ml)}}$$

For example, if the antigen was diluted 1:6 and after titration the first tube indicating complete fixation is 0.1 ml, then

$$\frac{6}{2 \times 0.1} = \frac{X}{0.5\,\text{ml}}$$

Thus, the antigen must be diluted 1 part plus 14 parts veronal buffer in order to contain 2 units per 0.5 ml.

It has been observed that many antigens are anticomplementary when tested in the presence of the hemolytic system. If they are not anticomplementary in the presence of negative serum (tubes No. 10 through 15), however, the antigen is considered satisfactory for use in the test proper. Some antigens are observed to be moderately turbid and, thus, do not afford a sparkling hemolysis even when all the blood cells are lysed. In some cases, extreme care must be taken at the time of reading in order to distinguish between turbidity due to the antigen from that due to the hemolyzed cells.

3. The Test Proper

a.   MATERIALS

> To perform the test proper, standardized reagents (see above) must be used:
> Antigen diluted to contain 2 exact units in 0.5 ml, cold.
> Complement diluted to contain 2 exact units in 0.5 ml, cold.
> Amboceptor diluted to contain 2 exact units in 0.5 ml, cold.
> Sheep red blood cells diluted to 2% concentration in veronal buffer, cold.
> Standard positive and negative sera for controls, cold.
> Test serum.
> Test tubes, pipets, water bath.

b.   METHOD

> (1) Pipet 0.1 ml of serum into each of two tubes containing veronal buffer in the amounts shown in Table 5-6. One will serve as the serum control. Place the tubes in a 56°C water bath for 30 min to destroy any complement native to the serum as well as to inactivate any anticomplementary factors present. Remove the rack from the water bath and allow the inactivated sera to stand at room temperature for 90 min.
> (2) Set up the protocol for the test proper as shown in Table 5-6.
> (3) Remove the rack from the water bath at the end of 45 min of incubation and read. To facilitate reading the test, the tubes may be

TABLE 5-6

The Test Proper

Reagents	Tube No. 1	Tube No. 2*
Buffer (ml)	0.4	0.9
Test serum (ml)	0.1	0.1
	Incubate at 56°C, 30 min, followed by 90 min at room temperature	
Add:		
Antigen, 2 units (ml)	0.5	—
Complement, 2 units (ml)	0.5	0.5
	Incubate at 37°C, 1 hr, or overnight at 4 to 6°C	
Add:		
Sensitized RBC (ml) (amboceptor + RBC)	1.0	1.0
	Incubate at 37°C, 45 min	

*Serum control.

centrifuged for a short time and then read, or the entire rack may immediately be placed in the refrigerator and the test read the next morning.

C. INTERPRETATION OF THE TEST PROPER

All serum controls (those tubes which did not receive antigen) should be completely hemolyzed. If any serum control tube contains non-hemolyzed cells, the sample is reported "anticomplementary" and no further interpretation is made as to the status of the animal's serum. Those tubes which contain the test samples plus antigen may show reactions from complete hemolysis (negative reaction) to complete fixation (positive reaction). The interpretation of those reactions falling between these two points is as follows:

4+	(no hemolysis)	positive
3+	(25% hemolysis)	suspicious
2+	(50% hemolysis)	suspicious
1+	(75% hemolysis)	suspicious
Trace	(only a few cells remaining)	negative
0	(complete hemolysis)	negative

4. Serum Titration
(Antibody Level Demonstration)

The titer of a positive serum is indicative of the stage of the disease whether it is from an acute or carrier animal. The test is performed by serially diluting the serum (starting with a 1 : 5 dilution) in buffer until a final dilution of 1 : 1280 is made. The reagents specified under the test proper, p. 114, are exactly the same except for the test serum which is diluted.

- a. Dilute the test serum 1 : 5 by adding 1.0 ml of the serum to 4.0 ml of veronal buffer. Inactivate it in the 56°C water bath for 30 min.
- b. Place 13 tubes in a rack and label according to the protocol in Table 5-7.
- c. Leave tubes No. 1 and 10 empty; to the others, add 0.5 ml veronal buffer.
- d. Add 0.5 ml of the inactivated serum to tubes No. 1, 2, and 10. Tubes No. 1 and 10 (serum controls) thus contain only serum at this point.
- e. Tube No. 2 now contains 1.0 ml of a 1 : 10 dilution. Mix thoroughly and transfer 0.5 ml to tube No. 3. Repeat this serial dilution until 0.5 ml has been removed from tube No. 9. Discard the last 0.5 ml. All tubes should now contain 0.5 ml of the respective dilutions.
- f. Add the standardized antigen and complement in the order shown in Table 5-7.
- g. Place the rack of tubes in a 37°C water bath for 1 hr, or overnight at 4 to 6°C followed by 10 min at 37°C.
- h. Remove. Add sensitized sheep red blood cells according to the protocol in Table 5-7. Incubate at 37°C for 45 min.
- i. Read the tests. To facilitate the reading of the titration end point (last tube showing a 4+ reaction), the tubes may be centrifuged for 5 min at 205 g or the entire rack may immediately be placed in the refrigerator and the tests read the following morning.

The serum titer is defined as the highest dilution showing a complete fixation (4+) reaction. Thus, if tubes No. 1 through 5 had 4+ reactions and tube No. 6 gave a 2+ reaction, the titer of the serum would be 1 : 80. The controls should read as follows:

Tube Number	Reaction
10 (serum control)	hemolysis
11 (amboceptor control)	hemolysis
12 (red cell control)	no hemolysis
13 (antigen control)	hemolysis

TABLE 5-7
Serum Titration

	Tube numbers and dilutions												
Reagents	1 1:5	2 1:10	3 1:20	4 1:40	5 1:80	6 1:160	7 1:320	8 1:640	9 1:1280	10* 1:5	11†	12‡	13§
Veronal buffer (ml)	—	0.5	0.5							—	0.5	1.0	—
Inactivated serum (ml)	0.5	0.5	0.5	0.5	0.5	0.5	0.5	0.5	0.5	0.5	—	—	—
Antigen, 2 units (ml)	0.5	0.5	0.5	0.5	0.5	0.5	0.5	0.5	0.5	—	—	—	0.5
Complement, 2 units (ml)	0.5	0.5	0.5	0.5	0.5	0.5	0.5	0.5	0.5	0.5	0.5	—	0.5

(Serial dilution from tube 2 through tube 9; 0.5 Discard)

Incubate for 1 hr in a 37°C water bath; or overnight at 4 to 6°C

| Amboceptor and RBC suspension (sensitized sheep RBC's) (ml) | 1.0 | 1.0 | 1.0 | 1.0 | 1.0 | 1.0 | 1.0 | 1.0 | 1.0 | 1.0 | 1.0 | 1.0 | 1.0 |

Incubate at 37°C for 45 min

*Serum Control
†Amboceptor Control
‡RBC Control
§Antigen Control

117

Each time titrations of unknown sera are performed, a standard positive serum having a known antibody titer should be included as a control.

5. The Reagent Test

If the unit values for complement, antigen, and hemolysin have been previously standardized, and large pools of the reagents are kept at -40 to $-70°C$, it may not be necessary to reestablish these units if tests are performed daily on a routine basis. A simple check of complement and hemolytic activity may be made in the following manner.

a. MATERIALS

(1) Complement, 2 exact units in 0.5 ml, cold.
(2) Amboceptor, 2 exact units in 0.5 ml, cold.
(3) Sheep red blood cells, diluted to 2% in veronal buffer, cold.
(4) Test tubes, pipets, rack.

b. METHOD

(1) The protocol for the complement and hemolytic check test is shown in Table 5-8. Place the rack of tubes in a 37°C water bath for 45 min.
(2) Read the tests as follows:
    (a) Tubes No. 1 and 2 will start to show hemolysis within a few minutes after placing the rack in the water bath.
    (b) Tube No. 3, which contains 1 exact unit of complement, will be equivalent to the amount of complement present in 1 exact unit

TABLE 5-8

The Reagent Test

Reagent	Tube Number			
	1	*2*	*3*	*4*
Diluted complement (ml)	0.5	0.375	0.25	0.125
Veronal buffer (ml)	1.0	1.0	1.0	1.0
RBC + amboceptor (ml)	1.0	1.0	1.0	1.0
Place in 37°C water bath, 45 min				
Units of complement contained in the quantities of diluted complement above	2.0	1.5	1.0	0.5

of a 1:25 dilution of complement as determined in the initial complement titration and should clear in 20 to 30 min.

(c) Tube No. 4 will read a 2+ or 3+ reaction at the end of 45 min. If you have correctly diluted the complement to contain 2 exact units per 0.5 ml, and tube No. 3 clears in the stated time and tube No. 4 does not hemolyze in 45 min, you may assume that all reagents are in the proper concentration and may still be used safely in the standard test.

G. BLOOD DISC ELUTION TECHNIQUE

On occasion, you must collect serum samples under field conditions. Needles, syringes, and test tubes might be inconvenient, however, to carry along. KARSTAD *et al.* [27] developed a technique whereby blood specimens are absorbed into filter paper discs. These require a minimum of space and equipment, are light weight, and easy to work with.

1. Materials

> No. 740-E antibiotic assay discs, Schleicher and Schuell
> Hanks' medium or PBS, sterile
> Test tubes, 12 × 72 mm, sterile
> Pipets, 1.0 and 10.0 ml, sterile
> Test tube rack
> Envelopes, small, cellophane

2. Method

a. Hold a disc with forceps and lightly touch the blood until the disc is saturated. Place the disc in a small envelope. Use one envelope for all discs from a single specimen. Label. These may be stored at room temperature until tests are made.

b. Place a disc in the bottom of a small test tube. Add 1.96 ml of diluent (Hanks' medium, PBS).

c. Place a 1.0 ml pipet on top of the disc to hold the disc submerged. Close the tube with the sterile cotton plug (Fig. 5-8).

d. Leave the disc submerged in diluent at room temperature for approximately 16 hr. The eluate is considered to approximate a 1:10 dilution (it is estimated that the disc absorbs 0.4 ml of blood).

e. Remove the disc with sterile forceps, using sterile technique.

f. Use the eluate as you would a 1:10 serum sample and test for antibody activity via the *alpha* or *beta* neutralization technique.

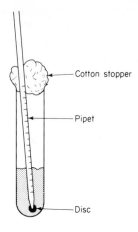

Fig. 5-8. Disc in test tube with pipet.

H. INTERSECTING SERUM SCHEME

If you have a large number of antisera to test, you can do so conveniently by preparing a pool scheme. This technique was developed by SCHMIDT *et al.* [49].

1. Materials

>   Antisera
>   10 or 100 TCID$_{50}$ virus
>   PBS, sterile
>   Pipets, 1.0 and 10.0 ml, sterile
>   Test tubes, 12 × 72 mm, sterile
>   Tubes of tissue culture, three for each serum pool
>   Test tube racks
>   Tissue culture racks
>   Water bath, 56°C

2. Method

a. Heat antisera at 56°C for 30 min. Dispense the antisera according to an intersecting scheme:

EXAMPLE

Pool	No. 1	No. 2	No. 3	No. 4
No. 5	S-1	S-2	S-3	S-4
No. 6	S-5	S-6	S-7	S-8
No. 7	S-9	S-10	S-11	S-12
No. 8	S-13	S-14	S-15	S-16

Note that a single serum appears in only two pools; i.e., serum S-1 appears only in pools No. 1 and No. 5; serum S-11 appears only in pools No. 3 and 7.

Note also that a single pool contains four antisera; i.e., pool No. 1 contains antisera S-1, S-5, S-9, and S-13; pool No. 7 contains antisera S-9, S-10, S-11, and S-12.

b. Add PBS to each pool to obtain a final dilution of each serum of 1:20.

EXAMPLE

Pool No. 5 has 0.1 ml of serum S-1
0.1 ml of serum S-2
0.1 ml of serum S-3
0.1 ml of serum S-4

Total 0.4 ml
Add: 19.6 ml PBS
∴   Serum S-1 is diluted 1:20
S-2 is diluted 1:20, etc.

c. Place equal volumes of an immune serum pool plus 10 and/or 100 $TCID_{50}$ of virus in small test tubes and incubate at appropriate temperature for the virus for 1 hr.
d. Add 0.2-ml samples of each serum-virus pool to replicate tissue cultures (Fig. 5-9):
e. Incubate cultures at 37°C, and observe them microscopically for 7 days.
f. Determine which, if any, serum neutralizes the virus.

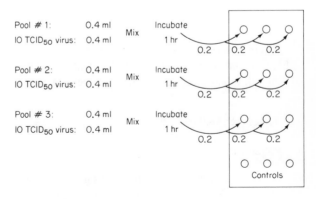

Fig. 5-9. Intersecting serum test.

Neutralization of virus X is noted in those tubes containing pools No. 2 and 6.

The only antiserum common to pools No. 2 and 6 is serum S-6.

Therefore, serum S-6 at a dilution of 1:20 neutralizes 10 or 100 $TCID_{50}$ of virus X and you may wish to perform more detailed neutralization tests with this particular serum.

I. FLUORESCENT ANTIBODY STAINING

Basically, the principle of immunofluorescent staining is simple. Immune globulin, separated from serum and concentrated, is conjugated (combined) with a stable fluorescent dye to produce a labeled antibody. The labeled antibody is used to detect the corresponding specific particulate antigen (bacteria, virus) in tissues or specimens, or a specific soluble antigen.

Four different methods are possible: indirect, direct, inhibition, and indirect-complement. The first two will be described here.

In the indirect method, antiglobulin serum, prepared in rabbits, is labeled with fluorescent dye. These "tagged" antisera are normally commercially prepared. Unlabeled serum is incubated with a known antigen and if the serum contains antibody specific for the known antigen, the serum will combine with it. If observed with the fluorescent microscope (which uses ultraviolet and blue light) at this point, no fluorescence will be seen. Labeled antiglobulin serum is then added. It combines with the antigen-antibody complex and, under the fluorescent microscope, the reaction is seen as a yellowish-green light (fluorescein conjugate) or reddish-orange (rhodamine conjugate) light. This indirect method can be used to determine the presence of virus in cultures where they cannot be detected by any other procedure.

In the direct method, specific antibody is conjugated with a fluorescent dye. Antigen is allowed to react with this labeled antibody and, when observed with the fluorescent microscope, a brilliant glow will be seen.

There is no one method which can be used for all viruses. The technique of fixation and staining is complex and will vary with the system under study so you will have to adjust the method accordingly.

1. Indirect Method

a. MATERIALS

Infected tissue culture cells grown on cover slips
PBS
Acetone

Unlabeled antiserum
Fluorescent-labeled antiglobulin
Glycerin, 9:1 in PBS
Staining jars
Petri dishes
Slides, 1 × 3 in.
Forceps
Long, narrow spatula
Gauze
Microscope equipped with a dark-field stop-iris diaphragm
 and a barrier filter, plus a 200-watt mercury arc illuminator
 with an exciter filter

b. METHOD

(1) Carefully remove cover slips from the tissue culture tubes by using a long, narrow spatula as described in Chapter 7, Section B.
(2) Grasp a corner of the cover slip with fine forceps and place it into a micro-carry-all (Chapter 9, Section K), held in a staining dish containing PBS (Fig. 5-10).

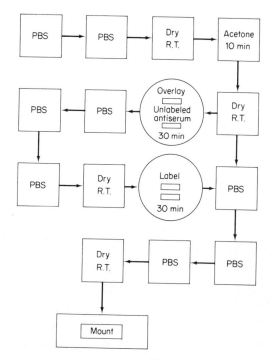

Fig. 5-10. Indirect fluorescent antibody staining procedure.

(3) Wash the cover slips by transferring the carry-all to another dish of PBS. Place the cover slips in an empty dish and allow them to dry at room temperature.

(4) Fix the tissues by transferring the cover slips to a dish containing acetone, and leave them for 10 min. Remove the cover slips to an empty dish and dry them at room temperature.

(5) Place a layer of moist gauze or filter paper in the bottom of a Petri dish. Carefully remove the cover slips from the carry-all and place them on the gauze.

(6) Overlay each cover slip carefully with unlabeled antiserum. Be conservative with serum as it is too expensive to allow it to overflow the cover slip. Incubate these at room temperature for 30 min.

(7) Carefully remove each cover slip with forceps and wash off the serum by dipping each cover slip in three containers of PBS. Place the cover slips in a dish and allow them to dry at room temperature.

(8) Carefully overlay each cover slip with labeled antiglobulin and leave them under a Petri dish cover for 30 min to permit conjugation with antiserum.

(9) Carefully wash each cover slip in three changes of PBS to remove excess labeled antiserum. Dry at room temperature.

(10) Mount by inverting each cover slip onto a clean 1 × 3 in. glass slide using glycerin (9:1) in PBS as the mounting medium.

2. Direct Method

a. MATERIALS

As indirect method, above.

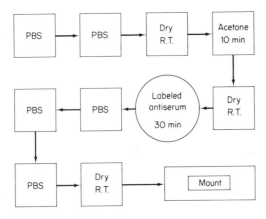

Fig. 5-11. Direct fluorescent antibody staining procedure.

b.   METHOD

(1) Follow the indirect method above through step 5 (Fig. 5-11).
(2) Carefully overlay each cover slip with labeled antiserum and incubate at room temperature for 30 min.
(3) Carefully wash each cover slip in three changes of PBS. Dry at room temperature.
(4) Mount in glycerin.

Biochemical and
Biophysical Characterization
of a Virus
SIX

There are many biochemical and biophysical tests which can be used as aids in characterizing and classifying viruses (Table 6-1). A few of the less complex ones will be discussed.

Choose a particular virus and use only that one for all the subsequent studies in order to define the virus's characteristics. This will afford an opportunity to develop a few techniques along with some understanding of the unseen things going on at the cellular and molecular levels.

A.   PLAQUE FORMATION

Cytopathic viruses (those which cause cell destruction) form plaques, foci, or local lesions in or on various indicator systems. No single method can be used to plaque-assay all animal viruses, and a method for each is beyond the scope of this text. The virus, cell strain, etc., will vary from one facility to the next so adaptations will have to be made with the methods discussed below. Lesions or pocks may be observed on the chorioallantoic membrane of the embryonating egg (Chapter 2), pustules may be produced on skin or cornea of various animals, foci or proliferation of tumors may be observed on cell monolayers, and plaque counts may be made on phage-infected bacterial cultures (Chapter 8).

Two methods exist for counting infective particles: end point (all-or-none response) and foci counting. You either score for growth or count foci, re-

TABLE 6-1

Criteria for Classification of Some Animal Viruses

Core (nucleic acid)	Capsid symmetry	Presence envelope	Size (nm)	Ether	pH 3.0	Heat (56°C)	Cationic stabilization to heat	CPE	Inclusions	Virus
DNA	Cubic	+	180–250	Sens.*	Sens.	Sens.	Sens.	Syncytial	Nuclear, central, eosinophilic with halo, Feulgen +	Herpes
		−	40–50	Res.†	Res.			Ballooning, vacuolization		Polyoma Papilloma (ϕX-174)
		−	60–85	Res.	Res.	Sens.	Sens., and Res.	Lysis	Nuclear, small, eosinophilic, Feulgen +	Adenovirus
	Cubic, tailed	−								(T-even phages)
	Helical	+	200–250			Res.			Cytoplasmic, Feulgen +	Poxvirus
	Cubic	+	15–120	Sens.	Sens.	Sens.	Sens.	Various	Various	Arbovirus
		−	< 50	Res.	Sens.					Rhinovirus
					Res.	Sens.	Res.	Rapid lysis	Nuclear and/or cytoplasmic, small with dense granules	Enterovirus
RNA			70–85		Res.	Res.	Res.		Cytoplasmic	Reovirus
	Helical	+	80–100	Sens.					Cytoplasmic	Influenza Fowl Plague
			60–125	Res.					Nuclear, cytoplasmic	Measles
		−								(Plant viruses)

*Sensitive
†Resistant

127

spectively. It is the latter which is used in plaque studies. The number of plaques is proportional to the amount of virus plated.

1. Chorioallantoic Membrane Pock Assay

A useful host for production of localized lesions is the chorioallantoic membrane of the embryonating egg (Chapter 2). The virus is serially diluted and deposited on the membrane of replicate eggs. The membrane, of uniform cells, is moist, and the viruses released can spread by cell contact. The characteristic lesions produced by various viruses are of diagnostic value.

a. MATERIALS

> Infected embryonating eggs
> Forceps, sterile
> Scissors, sterile
> Petri dish, sterile
> PBS, sterile
> Alcohol, 70%

b. METHODS

(1) Three to seven days after CAM inoculation, harvest the membranes.
  (a) Method 1: Disinfect the surface of the egg over the injection site. With sterile forceps, pick away the shell to the margin of the false air space. Cut out the membrane with sterile scissors. Place it in a Petri dish containing a few milliliters of PBS. Spread the membrane out so that it lies as flat as possible.
  (b) Method 2: Disinfect the area over the original air space (blunt end). With sterile forceps, remove the shell over this space. Cut a circular area in the shell membrane and in the CAM and fold this back. Tilt the egg and allow the contents to flow slowly out into a dish. Use sterile forceps to control the CAM, which should remain adherent to the inner side of the shell. Cut the embryo and attached membranes from the CAM. With sterile forceps, remove the CAM and place it in a Petri dish containing PBS.
(2) Place the dish against a dark surface and count the pocks. Some are minute and may require the aid of a dissecting microscope.

2. Monolayer Plaque Assay

Plaques can be seen more easily if an infected monolayer of cells in culture is overlaid with nutrient agar [13]. After several days, plaques will be seen as unstained areas in the cell sheet. At the viral concentration used, each plaque can be said to be caused by a single virus particle.

There are several important points you should realize when plaque work is attempted. The volume of the overlay can be critical. Some viruses require oxygen [3]; a thick overlay will reduce the number of plaques. Other viruses may require more nutrients, and their number may be reduced with a thin overlay.

The media used will depend on the cell culture and the virus. Some viruses require more elaborate nutrients [53, 58] so the medium suggested below may have to be changed. If bottle cultures are retained for a long period of time, nutrients must be added at intervals.

Serum in the overlay medium may inhibit or enhance plaque formation. If inhibition occurs, do not use serum, or try serum from a different species of animal.

There may also be inhibitors in the agar itself. Washing the agar, or using a more purified agar, may solve this problem. Clotted plasma [29], DEAE dextran [30], or 1.5% methyl cellulose [24] may be used as a substitute for the agar if necessary.

The volume of the inoculum will vary with the virus, dish, or bottle. If the surface of the bottles is not flat, the virus will be distributed unevenly or in a meniscus around the edge of the sheet while the center of the sheet will remain virus-free. A larger inoculum will be required.

It is necessary to avoid photoinactivation [20] of viruses which may occur due to the presence of neutral red in the overlay medium. Work in a darker room than usual or cover the bottles while the overlay hardens or use the overlay without incorporation of stain [36]. After the plaques have formed, you then add 1 or 2 ml of stain to each bottle and incubate at 37°C for 1 or 2 hr.

Cells are grown in bottles or dishes, medium is removed, and a suitable dilution of virus is added in small volume. After an adsorption period, nutrient agar is added and allowed to solidify. Cultures are incubated for varying lengths of time. Plaques may be seen by indirect light against a dark background.

a. MATERIALS

> Virus suspension
> Hanks' 10X medium without phenol red, sterile
> LAH, 5.0%, sterile
> Calf or other serum, sterile
> Neutral red, 1:1000, sterile
> Sodium bicarbonate (NaHCO$_3$), 7.5%, sterile
> Antibiotic mixture

Nobel agar, sterile
Distilled water, sterile
Erlenmeyer flasks, 100 and 200 ml, sterile
Tissue cultures in 3-oz prescription bottles, three per dilution
Pipets, 1.0 and 10.0 ml, sterile
Cornwall syringe and 18-gauge needle, sterile
Bunsen burner
Flat storage trays
Water baths, 37 and 45 to 48°C

b. METHOD

(1) Prepare medium A:

Hanks' 10X medium without phenol	10.0 ml
LAH, 5%	10.0
Calf serum	amount desired
Neutral red, 1:1000	2.4
NaHCO$_3$, 7.5%	0.7
Antibiotic mixture	0.1
Distilled water to bring volume to	60.0

Seal with a rubber stopper and place in a 37°C water bath.

(2) Prepare medium B:

Nobel agar	1.5 g
Distilled water to bring volume to	40.0 ml

Place in a 100 ml flask and cover with foil. Autoclave at 10 lb pressure, 10 min. Put the flask in a 45 to 48°C water bath and allow the contents to cool to that temperature for 1 hr (do not use a lower temperature because the agar will solidify).

(3) In the meantime, thaw the virus rapidly and make serial dilutions in Hanks' medium.

(4) Inoculate the tissue cultures as follows:

(a) With sterile technique, pour off the supernatant maintenance media into a discard container.

(b) Dispense 0.2 ml of virus dilution into replicate bottles. Manually rotate the inoculum over the surface of the cell sheet and allow the inoculum to remain in contact with the monolayer for 1 hr at the appropriate temperature for the virus. Do not permit cells to dry or be exposed to bright light during this period.

(5) At the end of the incubation period, set up your work area with the bottles of culture lined up on a tray, the Bunsen burner set in a convenient central spot, and the water bath holding the melted and cooled agar close by to your right.

(6) With sterile technique, combine Medium A and Medium B. Place

the container in the 45 to 48°C water bath. Working quickly, and with sterile technique, insert the metal ball on the Cornwall syringe into the agar medium. Gently fill the syringe to avoid massive formation of bubbles.

(7) With sterile technique, add 10.0 ml of agar medium to the first culture bottle. Do this by allowing the agar to flow down the side of the bottle opposite the monolayer. Replace the stopper and gently rotate the bottle, allowing the agar to flow over the cell sheet. Do not be vigorous with this procedure in order to avoid bubbles in the agar.

(8) Place the bottle down flat on the tray, with the agar covering the monolayer. Repeat for each bottle in turn. Allow them to remain undisturbed in a darkened room (or covered) until the agar is set. This will take approximately 1 hr. Do not attempt to invert the bottles any sooner as the agar is soft and though it gives the appearance that it is set, an early inversion will prove that it is not.

(9) Invert the bottles on the tray. This will prevent condensation from falling on the agar, causing a spread of virus over the monolayer with loss of distinct plaques. Incubate at 37°C.

(10) Examine daily for plaques. These will appear as "holes" in the agar. Some will be clear, others opaque or translucent. Some will have smoothly defined edges, others will have an irregular outline. Some will be large, others small. A particular virus will produce a particular plaque type (Fig. 6-1).

Fig. 6-1. Plaques in cell monolayers.

3. Agar Suspension Plaque Assay

Nutrient agar is added to the bottles and allowed to solidify first, as a base layer. These may be prepared in advance (1 or 2 weeks) and stored at 4°C. An appropriate viral dilution in a small volume is mixed with liquid agar-cell suspension, and poured over the base layer. This procedure [7, 8] allows for greater plating efficiency, eliminates the toxic effects of contact of virus with glass surfaces, and has the advantage that no adsorption times are needed. In some cases (ascites tumor cells, for example) it is the only method of plaque assay, as such cells do not form monolayers.

a. MATERIALS

> Virus dilution containing 10 to 100 PFU/ml
> Agar-medium suspension, above, sterile
> Hanks' maintenance medium, sterile
> Trypsinized tissue culture cells
> Petri dishes, sterile, 100 mm, or bottles, sterile
> Test tubes, 12 × 72 mm, sterile
> Pipets, 1.0 and 2.0 ml, sterile
> Water bath, 45 to 48°C

b. METHOD

(1) Melt and cool nutrient agar Medium B and mix with Medium A as discussed above. Pour 10.0 ml into each dish or bottle. Allow to solidify. Hold remaining agar at 45 to 48°C.

(2) Prepare trypsinized tissue cells in maintenance medium and suspend at 1.0 or 1.5×10^8 cells per milliliter. Place 1.5 to 2.0 ml of cell suspension into small sterile test tubes.

(3) Prepare serial dilutions of virus in maintenance medium. Add 0.1 ml of each virus dilution to a tube of cell suspension. Add 2.0 ml of cooled agar. Quickly mix and pour onto the base layer.

(4) Allow the second layer to solidify in darkness; then incubate as required. Count as usual.

4. Plaque Counting and Determination of Titer

Foci are usually counted with the unaided eye, although some may have to be checked microscopically. Pocks on CAMs should be observed against a dark background or oblique illumination from below may be used. Try to distinguish non-specific lesions which usually occur in the area of the original injection site, or along veins, in groups or irregular clumps. If too high a concentration of virus was used, pocks may become confluent. This may also occur if you did not rotate the egg after inoculation. There may be edema and hemorrhage present. Carefully compare test inoculates with control CAMs prepared and harvested simultaneously.

Hold plates or bottles (not inverted) against a dark background and mark each plaque with a glass marking pen to avoid duplication of counts. Avoid gaps or tears in cell sheets. Plaques will normally appear gray and should be distinguished from a torn monolayer.

   a. Count in that bottle, dish, or on that CAM which shows a convenient number of non-overlapping plaques. A good rule of thumb is to count those with 30 to 100 plaques.
   b. The titer is calculated by the formula

$$\frac{\text{Average number of plaques}}{\text{Volume of inoculum} \times \text{dilution}} = \begin{array}{l}\text{number of plaque-}\\\text{forming units (PFU)}\\\text{per milliliter of}\\\text{original suspension}\end{array}$$

EXAMPLE

One-tenth milliliter of a 10^{-5} dilution of virus suspension was inoculated into replicate bottles. Two and five plaques appeared in the two bottles (average, 3.5). The titer may be calculated:

$$\frac{3.5}{0.1 \times 10^{-5}} = \begin{array}{c}3.5 \times 10^{-5}/0.1 \text{ ml}\\\text{or}\\35 \times 10^{-5}/1.0 \text{ ml}\end{array}$$

The dishes at the next lower dilution (10^{-4}) would contain about 10 times as many plaque-forming units (about 35 per bottle) and would be difficult to count due to overlapping plaques. The next higher dilution (10^{-6}) would contain less than one plaque (i.e., you would expect none) per bottle.

5. Picking From Plaques

On occasion you may wish to harvest ("pick") virus from the plaque areas. If you have well-isolated plaques, this is a good method of harvesting a "pure" virus. To assure that the plaque does not have multiple origins, it is technically better to harvest from indicator systems with one to several plaques.

a.   CHORIOALLANTOIC MEMBRANE

   (1) *Materials*

              Tissue containing pocks
              Forceps, sterile

Scissors, sterile
Ten Broeck tissue grinder, sterile
Hanks' maintenance medium, sterile
Pipets, 1.0 and 10.0 ml, sterile
Vials, screw-capped, sterile
Ice bath

(2) *Method*

(a) With sterile forceps and scissors, cut the tissue around a pock and remove the pock to a Ten Broeck grinder held in an ice bath.

(b) Add a minimum measured volume of medium and homogenize the tissue. Clear by centrifugation.

(c) Transfer the supernate to a sterile vial. Label, date, and store as required.

b. MONOLAYER [25]

(1) *Materials*

Infected tissue cultures showing plaque formation
Uninoculated tubes of tissue culture
Pasteur pipets, cotton plugged, sterile
Rubber bulb
Bunsen burner

(2) *Method*

(a) Hold the tapered end of a Pasteur pipet in a Bunsen flame just long enough to allow the last half inch to heat up and bend at almost a right angle [Fig. 6-2(a)]. Be sure the opening has not been occluded in the area of the bend. Allow the end of the pipet to cool while you put on the rubber bulb.

(b) With sterile technique, open the bottle containing the tissue cultures showing plaque formation. Depress the bulb on the pipet, insert the tip into the bottle, and (with a steady hand) place the tip of the pipet into the center of a plaque [Fig. 6-2(b)].

(c) Gradually release the bulb and allow the material from the center of the plaque to be sucked into the tip of the pipet [Fig. 6-2(c)]. Do not release the bulb entirely or the plaque material will be sucked high up into the pipet.

(d) Remove the pipet and expel its contents into the fluid contained in an uninoculated tube of tissue culture [Fig. 6-2(d)].

(e) Repeat with a new sterile Pasteur pipet and a new tube of uninfected tissue culture for each plaque you wish to pick. Incubate the tissue cultures at 37°C, check for CPE, and harvest as usual. Indicate plaque passage on the label. Freeze.

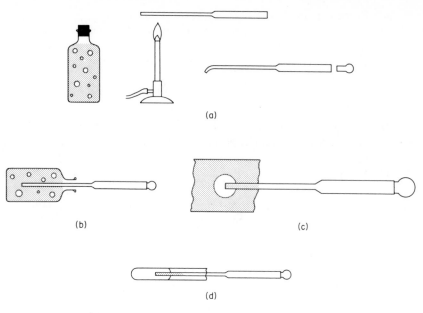

Fig. 6-2. Plaque picking.

6. Plaque Reduction (Neutralization) Method

Virus inhibitors (antibody, for example) can be estimated by this method.

a. MATERIALS

> As described above (2a)
> Positive and negative control sera
> Unknown sera
> Test tubes, 12 × 72 mm, sterile
> Water bath, 56°C

b. METHOD

(1) Determine the virus titer by the plaque agar-overlay method, above. Prepare a constant virus dilution in Hanks' medium to contain approximately 50 to 100 PFU per 0.1 ml.

(2) Heat inactivate (56°C, 30 min) an aliquot of all serum samples: virus-specific (positive), normal (expected negative), and unknowns (those you are testing). Use sera which are either undiluted or diluted (see C, below).

(3) Add an equal volume of the serum to be tested to an equal volume of the selected virus dilution(s). See the example, Section D, below.

(4) Prepare a standard virus control titration.

(5) Incubate the virus control titration and the virus-serum mixtures at room temperature for 1 hr.

(6) Inoculate 0.2 ml of the virus-serum mixtures into replicate bottle cultures.

(7) Inoculate 0.1 ml of virus control dilutions into replicate bottle cultures.

(8) Lay bottles in trays so that the virus is in contact with the cell sheets and place at 37°C (or appropriate temperature) for 1 hr to allow for adsorption of virus.

(9) In the meantime, prepare and cool agar-overlay medium as described above.

(10) Remove the bottles from the incubator and carefully overlay the cell sheets with agar medium. Cover the bottles and allow the agar to solidify. Invert the bottles on trays and incubate at 37°C or appropriate temperature.

(11) Observe the cultures daily and count the number of plaques as they appear. Determine the titer of the virus alone and in combination with the negative serum. No neutralization should occur; if it does, there is something wrong with your test or technique, and you should recheck your procedures. The positive serum should neutralize the virus to the same degree as in previous tests you have made.

(12) Determine the neutralization index (N. I.) of the unknown samples by subtracting the endpoint of the virus, tested after being mixed with serum, from the endpoint of the virus control titer. For example, the virus control endpoint $= 10^{-5}$; the virus + serum endpoint $= 10^{-3}$; therefore, the N. I. is $10^{-5} - 10^{-3} = 10^{2}$. If the serum was diluted 1:4, then the N. I. would be 4×10^{2}.

EXAMPLE (Fig. 6-3)

A. *Preparation of Dilutions for Virus Titration*

   1. Prepare tenfold serial dilutions of virus in Hanks' medium (see Chapter 4, Section C).

   2. The final concentration of virus is

$$10^{-1} \quad 10^{-2} \quad 10^{-3} \quad 10^{-4} \quad 10^{-5} \quad 10^{-6} \quad 10^{-7} \quad 10^{-8}$$

   3. Anticipated PFU per 0.1 ml is

300–500 30–50 2–5   <1   none   none

It is presumed that a reference supply of virus had been prepared prior to attempting this test. Several determinations of the titer of the reference supply should have been made and, in this test, the

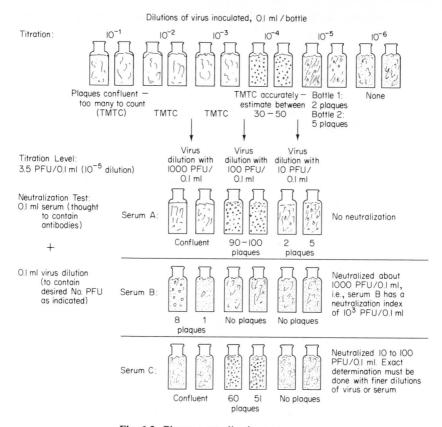

Fig. 6-3. Plaque neutralization test.

end point of the virus is expected to be like that of the previous tests. In this hypothetical example, the end point is expected to occur again at the 10^{-5} dilution. Therefore, for the neutralization test, you must select a dilution of virus which will give you the number of PFU per 0.1 ml which is to be mixed with the serum. Although, in doing a neutralization test, you will always include a titration of the virus as a check, you must "guess" the dilution (from previous titrations of the virus) which will *give* you the PFU per 0.1 ml you want.

B. *Preparing Virus Dilutions for the Neutralization Test*

1. You must select the number of plaque-forming units per amount of inoculum you wish to use in the neutralization test. For example, you may wish to test your serum against 10, 50, 100, etc., PFU. If your titer was done using 0.1-ml amounts per bottle, balance your dilution in the neutralization tests to the same amount; i.e., the titer of the virus was 3.5 PFU per *0.1 ml* of 10^{-5}. The serum samples in the neutra-

lization test will be tested against 10 PFU per *0.1 ml* of the appropriate dilution.

2. *Method*

    a. Select that level of virus titer dilution which is closest to the dilution you wish to make. From part A, such a titration dilution would be 10^{-4} (35 PFU per 0.1 ml) if you wish to make a dilution of 10 PFU per 0.1 ml for the neutralization test, or 10^{-3} (350 PFU per 0.1 ml) if you wish to make a dilution containing 100 PFU per 0.1 ml. You must now dilute the concentrated virus to give you the desired level to combine with your sera.

    b. Computation:

	To obtain 10 PFU per 0.1 ml	To obtain 50 PFU per 0.1 ml
(1) Select next most concentrated titer dilution:	10^{-4} (35 PFU)	10^{-3} (350 PFU)
(2) Determine the dilution necessary for the PFU wanted:	$\dfrac{35 \,(\text{number PFU in concentrate})}{10}$ $= 3.5$	$\dfrac{350}{50}$ $= 7$
(3) Make dilution:	1 part of 10^{-4} 2.5 parts diluent —————— 3.5 total dilution	1 part of 10^{-3} 6 parts diluent —————— 7 total dilution

C. *Preparing the Serum Dilutions*

    Usually, one selects a single serum dilution, and this is tested with the virus dilution prepared above. The following shows how to prepare several serum dilutions should it be desirable to test several levels simultaneously.

	Final Dilution of Serum:		
Materials:	1:2	1:4	1:10
Serum (ml)	0.5	0.25	0.1
Hanks' medium (ml)	0.5	0.75	0.9
	Mix. Discard pipet	Mix. Discard pipet	Mix. Discard pipet

D. *Test Proper* (where several levels of serum are tested against one level of virus*)

---

* Where only 1 level of serum is to be tested, the protocol still applies.

1. Prepare the serum-virus mixture and incubate at room temperature for $\frac{1}{2}$ to 1 hr:

Materials:	Serum Dilutions (from part C)		
	1:2	1:4	1:10
Serum (ml)	0.5	0.5	0.5
Virus (ml)	0.5	0.5	0.5

2. Inoculate 0.2 ml into each bottle. Be sure to use several for each dilution tested.
3. Incubate at 37°C, $\frac{1}{2}$ to 1 hr.
4. Overlay with agar medium. Cover, allow to solidify, and incubate at 37°C in an inverted position. Observe daily.

B. REPLICATION CYCLE

The determination of growth curves for the animal viruses is essentially the same as that first developed for bacterial viruses. A suspension of virus is added to a culture of cells and time is allowed for adsorption of the virus to the cell. The cell sheets are washed with buffer to remove unadsorbed "seed" virus, nutrient medium is added, and the cultures are incubated. At intervals thereafter the supernatant fluids are assayed for the presence of released virus, and cell sheets are disrupted to determine the presence of intracellular virus.

Following adsorption and penetration of the cell, a period of time occurs during which no detectable virus can be recovered. This is called the *eclipse period*. This is followed by a period of maturation of the virus. These two phases are collectively known as the *latent period*. The latent period for animal viruses is longer than that for bacterial viruses, the exact time depending on the particular virus-cell combination. For example, single-stranded RNA viruses, without envelopes, have a latent period of approximately 2 hr; myxoviruses, 3 hr; reoviruses, 7 hr; Rous sarcoma virus, 12 hr; adenovirus in KB cells, 14 hr.

The latent period is followed by a rise period during which virus is gradually released from the cell. The release of virus begins before the cell dies, the time varying with the virus-cell combination. Rous sarcoma virus in chick fibroblasts, for example, is released in less than 30 min; foot-and-mouth disease virus in 60 min; herpes in 3 hr; mengo virus in 8 to 12 hr. After cell death, much of the virus remains within the cell or attached to the debris. If a growth curve is drawn for a particular animal virus, it will be seen that the total virus (intracellular plus extracellular virus) curve is higher than the released virus (extracellular) curve.

1. Materials

> Virus suspension
> Uninoculated tissue cultures
> Hanks' maintenance medium, sterile
> PBS, sterile
> Centrifuge tubes, sterile
> Pipets, 1.0 and 10.0 ml, sterile
> Test tubes, 12 × 72 mm, sterile
> Tissue culture racks
> Test tube racks
> Water bath, 37°C
> Freezer

2. Method

a. Set up 13 rows of tissue culture tubes, 5 tubes in each row.

b. Prepare 100 $TCID_{50}$ of virus in Hanks' medium. Inoculate each culture tube with 0.1 ml of 100 $TCID_{50}$ virus. Incubate at 37°C.

c. At hourly intervals for 5 hr consecutively, remove 5 tubes of culture, pool the fluids, centrifuge to remove any cell debris, label, and freeze the supernatant fluid at −70°C. This will be the source of extracellular virus. (For those viruses sensitive to freezing, immediate titration in the proper indicator system will be required.)

d. Wash the cell sheets remaining in those 5 tubes at least five times with PBS, using sterile techniques. Pour the wash fluids into a contamination pan. Add 1.0 ml of Hanks' medium to each tube, label, and place them at −70°C. This will be the source of intracellular virus.

e. After 5 hr, you should have 8 rows of 5 tubes remaining. Wash the cell sheets thoroughly with PBS (i.e., pour off the infected fluids from the remaining 40 tubes). Add 1.0 ml of fresh maintenance medium to all tubes and reincubate the cultures.

f. At intervals of 8, 12, 24, 36, 48, 72, 96, and 120 hr (or at other times found to be convenient), remove 5 tubes and treat as in steps c and d above.

g. When the test is completed, freeze and thaw all tubes of intracellular and extracellular virus three times.

h. Prepare tenfold serial dilutions of all samples. Titrate each sample in tissue culture to determine intracellular and extracellular virus activity.

i. Draw a growth cycle diagram (Fig. 6-4) to indicate eclipse, latent period, maturation, and release phases for your virus by indi-

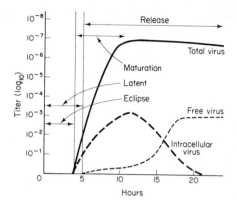

Fig. 6-4. Sample growth curve.

cating total virus, intracellular virus, and extracellular virus titer.

C. DETERMINATION OF SIZE

Several methods have been developed to aid in the determination of viral size. Direct methods of measurement involve the use of an electron microscope whereby size as well as shape can be determined. Electron microscopic techniques involve the use of specialized materials and equipment not usually available in the average laboratory. These methods will not be discussed here.

It is a common practice to measure indirectly viral size by techniques using untreated membrane filters. Virus suspensions are first clarified by passing them through 300 and 200 mμ filters. The clarified filtrates, passed through smaller porosity membranes, are used for determination of size. Viruses can be adsorbed to these membranes, however, even those whose porosity is two or three times greater than the diameter of the virus under study.

Ver *et al.* [55] developed a procedure by which virus suspensions are passed through membranes whose porosity is twice the diameter of the virus. The filtrates, containing 50 to 100% of the initial viruses present, are used for sizing virus by passage through smaller porosity membranes. Membranes are pretreated with protein (10% calf serum) which adsorbs to membranes and precludes virus adsorption. Viruses which are resistant to detergents can be treated with sodium lauryl sulfate and can be filtered through untreated membranes. Detergent-sensitive, lipophilic viruses can be diluted in distilled water to reduce the salt concentration (to below 0.002 M) and surface tension and then filtered through untreated membranes.

While size alone is not adequate for viral classification, it is an interesting parameter to the over-all study of viruses. The actual size of the virus may be considered to be approximately one and a half times smaller than the lowest filter through which it can pass. The estimation of size by filtration takes as a first presumption that the dimensions of the virus are uniform, i.e., spherical. Where this is true, the only limitation of a virus passing through the pores of a filter (other than adsorption to the filter) would be its size. Where a virus is not spherical, such as the pox viruses, many may not pass through the filter because they lie across and occlude the filter openings. The loss of virus numbers from such a test might be interpreted as an indication that the viruses being filtered are of a size which would require a larger porosity filter. Size determinations by filtration must be interpreted with care until the configuration of the viruses concerned is determined. The availability of an electron microscope will facilitate both configurational studies and size determinations and can be used in conjunction with, or as an alternate to, filtration studies (Fig. 6-5).

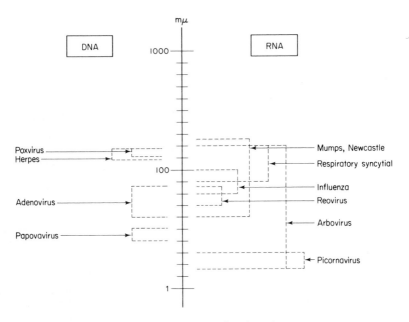

Fig. 6-5. Nomogram for virus sizes.

1. Materials

Virus suspension
Uninoculated tissue cultures
Hanks' medium, sterile
PBS, 10X, sterile

Distilled water, sterile
Sodium lauryl sulfate, 10%, sterile
Fetal calf serum, sterile
450, 300, 200, 100, 50 mμ membrane filters, 25-mm diameter, sterile
Clarifying pad, similar to Millipore type AP 25
TRIS buffer
Filter adapters, sterile
Syringes, sterile
Test tubes, sterile
Pipets, 1.0, 5.0, and 10.0 ml, sterile
Test tube rack

2. Method

a.   PREPARATION OF COATING PROTEIN

Prepare 100 ml of 10% calf serum in distilled water. To sterilize the serum, filter it through a Millipore AP 25 clarifying pad at 25 psi. Then pass the serum through 300, 200, 100, and 50 mμ membranes, in series, contained in a syringe adapter. Add 10X PBS to the final filtrate in such volume as to restore isotonicity.

b.   PRETREATMENT OF FILTERS

Treat membrane filters with 5 ml of calf serum prepared above by passing it through 25-mm membranes whose porosity is twice that expected for the virus to be tested.

c.   VIRUS FILTRATION, TREATED MEMBRANES

(1) Rapidly thaw the virus suspension. Centrifuge at 205 g for 10 min to remove cell debris. Remove one sample to be titrated as an unfiltered control. Treat other samples as indicated below.

(2) Prepare a filter holder containing a series of treated membranes (for example, 200, 100, 50 mμ) and attach it to a syringe.

(3) Filter 5 ml of clarified virus suspension through the membrane series at 1 psi and collect the filtrate in a sterile test tube.

(4) Titrate each filtered and unfiltered sample of virus in an appropriate indicator system and determine the titer.

(5) *Note:* If preferred, virus may be filtered through separate membranes of decreasing porosity. Remove a sample of each succeeding filtrate for virus titration.

d.   VIRUS FILTRATION, UNTREATED MEMBRANES

(1) Pass a 5-ml sample of undiluted virus through a single filter holder containing untreated membranes in series. Assay the filtrate in an appropriate indicator system.

(2) Dilute another sample of virus one hundredfold in distilled water and filter 5 ml through another series of untreated membranes. Assay the filtrate.

(3) Dilute another sample of virus (nine parts) with one part of a 10% solution of sodium lauryl sulfate. Filter 5.0 ml of this suspension through a series of untreated membranes. Assay the filtrate.

3. Determination of Size

If there has been no adsorption of virus to the filter membranes, there should be no, or only slight, decrease in titer of the filtrate (compared with the control) after passage through a certain porosity membrane and complete disappearance of virus after passage through the next smaller membrane. For example, the control titers at $10^{-6.5}$. The test virus suspension titers at $10^{-6.25}$ after passage through a 100 mμ filter and no virus is detected after passage of the suspension through the 50 mμ filter. It can be assumed, therefore, that the virus is at least 100 mμ, and probably somewhere between 50 and 100 mμ, in size.

If the control virus titers at $10^{-6.5}$ and the test virus suspension titers at $10^{-6.25}$ after passage through a 200 mμ membrane, $10^{-3.5}$ after passage through a 100 mμ membrane, and no virus can be detected after passage through a 50 mμ membrane, for example, two possibilities exist. Either some adsorption is still occurring or the shape of the virus is not spherical. Adsorption should be ruled out since little, if any, occurred with the 200 mμ membrane (adsorption should apply to all the membranes). It is more likely that the virus is actually somewhere between 50 and 100 mμ in size and is not spherical since it can pass through the 200 mμ membrane with no loss in titer but begins to drop sharply, but not completely, on passage through the 100 mμ porosity filter.

D.   EFFECT OF HEAT

In addition to its role as an aid in viral classification, you should know what effect heat will have on the particular virus you are using. You may want to destroy viral activity on some occasions (during autoclaving, for instance) and preserve viral activity (during storage) on other occasions. Most viruses are quite stable at room temperature when suspended in tissue or complex media. Measurable differences arise at temperatures from 50 to 60°C where protein denaturation is rapid. The standard test, below, uses 56°C as the constant temperature. There are differences among viruses, however. Some are stable at a higher temperature, whereas others are inactivated at temperatures much lower than 56°C.

To measure thermal inactivation, the virus suspension is held at a constant temperature. At intervals, samples are removed and assayed in an indicator system.

1. Materials

Virus suspension
Uninoculated tissue cultures or other indicator system
Hanks' maintenance medium
Test tubes, 12 × 72 mm, sterile
Pipets, 1.0 and 10.0 ml, sterile
Test tube rack
Tissue culture rack
Water bath, 56°C

2. Method

a. Place an undiluted suspension of virus in a small, sterile test tube. Place it in a 56°C water bath to a level equal to that of the suspension in the tube.

b. At intervals of $\frac{1}{2}$, 1, 2, 4, 6, and 8 hr (or other times, as desired), remove a sample and prepare serial dilutions of it in Hanks' medium.

c. Titrate as usual in tissue culture or other host system. Compare with an unheated virus control titered simultaneously. Loss of greater than 1 $\log_{10}$ in titer in 1 hr is considered to be indicative of heat lability.

d. Harvest and repass the heat-treated virus cultures. Remember that in a test which determines the lability of the CPE antigen, one is not necessarily determining the viability of the remaining viral genome. Viral antigens are separable from the infective particle and unless repassage of the treated material is performed, one can only state that the CPE antigen was labile or stable. For example, if the heat-treated material caused no CPE while the untreated virus produced CPE, a harvest and repassage of the cultures into which the treated material was inoculated is, then, necessary. Should CPE reappear on the first, second, or third serial passage, one can postulate that originally only the CPE antigen was destroyed but not the infectious moiety. When the virus was recovered and repassed, the intact replicated virus produced all required antigens.

E. SENSITIVITY TESTS WITH LIPID SOLVENTS

There appears to be a correlation between the presence of an envelope and susceptibility of viruses to lipid solvents. Although lipids are soluble in various organic solvents, viral characterization is commonly based on susceptibility to ether, chloroform, and formalin. Some viruses (herpes, myxovirus) are susceptible to these agents; others are resistant.

Ether acts by removing or fragmenting the viral membrane. It is commonly used to remove the membrane of myxoviruses and leaves RNA-protein cores exposed. The test is not of value for classification of poxviruses which behave variously to ether. Formaldehyde, which acts primarily on viral coat protein, is used in the preparation of vaccines because it does not affect viral antigens to the same degree.

1. Sensitivity to Ether

The method described is that of ANDREWES AND HORSTMANN [1].

a.   MATERIALS

> Virus suspension
> Uninfected tissue cultures or other indicator system
> Ethyl ether, pure, anaesthetic grade
> NaCl, 0.85%, sterile
> Two screw-capped test tubes, sterile
> Adhesive tape
> Two Petri dishes, sterile
> Pipets, 1.0 and 10.0 ml, sterile
> Test tubes, 12 × 72 mm, sterile
> Test tube rack
> Refrigerator, 4°C

b.   METHOD

(1) Place a measured amount of virus suspension into a screw-capped tube. Add 20% by volume of ether. Prepare a control virus suspension containing 20% by volume 0.85% NaCl. Close the tubes tightly

and tape with adhesive. Shake each tube manually. Place at 4°C for 18 to 24 hr, with manual shaking of the tubes at intervals.

(2) Pour each suspension into sterile Petri dishes, partially cover with the lids, and allow the ether to evaporate. Preferably, this should be done under a hood. Otherwise, be sure no flames are burning and that adequate ventilation is provided.

(3) Titrate the residual materials treated with ether and saline, and an untreated viral suspension as a control, in an appropriate indicator system. A drop in titer of 1 $\log_{10}$ is considered as susceptibility to the lipid solvent.

2. Sensitivity to Chloroform

The method of FELDMAN AND WANG [17] is used and is believed to give more reliable results than those obtained with ether.

a. MATERIALS

> Virus suspension
> Uninoculated tissue culture cells or other indicator system
> Hanks' medium
> Chloroform ($CHCl_3$), reagent grade
> Centrifuge tubes, sterile
> Pipets, 1.0 and 10.0 ml, sterile
> Tissue culture rack
> Test tubes, 12 × 72 mm, sterile
> Test tube rack

b. METHOD

(1) Centrifuge virus suspensions to remove cellular debris. To 1.0 ml of virus, add 0.5 ml of chloroform. To another 1.0 ml of virus suspension, add 0.5 ml of salt solution as a control. Shake both tubes manually for 10 min at room temperature, or mix using a tuberculin syringe fitted with a 26-gauge needle. Centrifuge at 33 g for 5 min.

(2) Use the uppermost, clear layer (bottom layer is chloroform, middle opaque layer is interphase) for viral titration. Titrate the salt-treated virus suspension. A drop in titer of 1 $\log_{10}$ is considered susceptibility to the lipid solvent.

3. Sensitivity to Formalin

For this test, the method of SALK, *et al.* [46] is described.

a.   MATERIALS

> Virus suspension
> Uninoculated tissue culture cells or other indicator system
> Hanks' medium, sterile
> Commercial formalin (37%), 1:2000 dilution
> Test tubes, 12 × 72 mm, sterile
> Pipets, 1.0 and 10.0 ml, sterile
> Test tube rack
> Tissue culture rack
> Water bath, 37°C

b.   METHOD

(1) Mix virus suspensions half and half with a 1:2000 dilution of formalin. As a control, use equal parts of virus suspension and Hanks'
medium. Incubate at 37°C. At intervals of 1, 2, 3, 4, 5, 6, and 24 hr,
remove a sample and dilute it 1:100 in Hanks' medium.

(2) Titrate the treated and untreated viral suspensions in the appropriate
indicator system. A drop in titer of 1 $\log_{10}$ is considered susceptibility
to the lipid solvent.

F.   pH SENSITIVITY

Viral sensitivity to acid pH can be tested according to KETLER et al. [28].

1. Materials

> Virus suspension
> Uninfected tissue culture cells or other indicator system
> Hanks' medium, sterile
> HCl, 0.1 N
> Test tubes, 12 × 72 mm, sterile
> Pipets, 1.0 and 10.0 ml, sterile
> Test tube rack
> Tissue culture rack

2. Method

a. Add dilute HCl to Hanks' medium, removing and testing samples
with a pH meter until a pH of 3.0 is reached. Do not return the
testing fluids to the bottle of sterile medium!

b. Dilute the virus sample 10^{-1} in pH 3.0 medium. Dilute another sample in Hanks' medium as a control. Incubate both at room temperature for one half, and one hour.

c. Prepare serial dilutions of each in Hanks' medium and titrate in the appropriate indicator system. A reduction in titer of 1 $\log_{10}$ or more after 30 min is interpreted as lability. Retention of full titer, or loss of less than 1 $\log_{10}$ of activity, is considered to be indicative of stability.

G. STABILIZATION BY CATIONS

It has been found [57, 59] that, when present in high concentration, monovalent and divalent cations appear to stabilize infectious RNA of certain viruses against heat. Enteroviruses (polio, Coxsackie, and ECHO) are of this type. Other viruses (myxoviruses, papova, arboviruses) are rapidly inactivated in the presence of divalent cations.

1. Materials

Virus suspension
Uninoculated tissue culture cells or other indicator system
Hanks' medium
Magnesium, 2 M, as $MgCl_2$
Calcium, 2 M, as $CaCl_2$
Sodium, 4 M, as NaCl
Test tubes, screw-capped, sterile
Test tubes, 12 × 72 mm, sterile
Pipets, 1.0 and 10.0 ml, sterile
Test tube racks
Tissue culture racks
Water baths, 37 and 56°C
Ice water bath
Refrigerator

2. Method

a. Prepare 20 ml of a tenfold dilution of virus in Hanks' medium. Place 1.0 ml into each of 16 small tubes. Add 1.0 ml of 2 M magnesium to each of 4 tubes; to another 4 tubes, add 1.0 ml of 2 M calcium; add 1.0 ml of 4 M sodium to the next 4 tubes (final dilution: 1 M Mg, 1 M Ca, 2 M Na) and 1.0 ml of Hanks' medium to the last 4 tubes as controls.

b. Place 1 tube from each set into a 37°C water bath and another set in a 56°C water bath to a depth within $\frac{1}{4}$ in. of the top of the tubes.

c. Place another set at room temperature (25 to 27°C) and the fourth set in a refrigerator at 4°C.

d. Tests performed at 56°C may be tested at $\frac{1}{2}$, 1, 2, 3, and 16 hr; 37°C tests after 24, 48, and 72 hr; room temperature tests after 1, 5, 10, 15, and 20 days; 4°C tests after 1, 2, and 3 months. Any other combination may be used as desired. At the end of the various test times, transfer the heat-treated samples to an ice bath to stop the action of heat. As each sample is removed, dilute it tenfold (1 : 100 dilution) in Hanks' medium to remove toxicity of the ions for cells. Make regular viral titrations of tests and controls in an appropriate indicator system. Incubate and read.

H. DETERMINATION OF NUCLEIC ACID TYPE

Structural analogs of naturally occurring metabolites possess selective virus-inhibiting activity [52]. FUDR (5-fluoro-2'-deoxyuridine) acts as a potent inhibitor of the enzyme thymidylate synthetase and consequently of DNA synthesis and is followed by a decrease in rates of RNA and protein synthesis [6]. Viruses inhibited by concentrations of approximately 10^{-5} M FUDR are likely to be DNA type [47]; those not inhibited can be classified as RNA-type viruses. BUDR (5-bromo-2'-deoxyuridine) and IUDR (5-iodo-2'-deoxyuridine) replace thymine in newly synthesized DNA, rendering it defective and mutagenic, but do not prevent synthesis of DNA [37]. Viruses can be divided into those which are susceptible to the action of halogenated pyrimidines and those that are not. Viruses whose genetic information is in the form of RNA are not affected by deoxyuridine analogs [2].

1. Materials

Virus suspension
Uninoculated tissue cultures
Hanks' maintenance medium
5-Fluoro-2'-deoxyuridine, 10^{-4} M
5-Bromo-2'-deoxyuridine, 50 μg/ml
5-Iodo-2'-deoxyuridine, 50 μg/ml
Test tubes, 12 × 72 mm, sterile
Pipets, 1.0 and 10.0 ml, sterile
Test tube racks
Tissue culture racks

2. Method

a. With sterile technique, discard the fluid from 30 tubes of tissue culture cells. Add 1.0 ml of pyrimidine-containing medium (i.e.,

a. 10^{-4} *M* solution of the analog in Hanks' medium) to each tissue culture tube.

b. Prepare a tenfold serial dilution of virus in Hanks' medium and inoculate triplicate pyrimidine-treated cultures with each dilution, 0.1 ml per tube. Inoculate triplicate pyrimidine-free control cultures with each dilution of virus, 0.1 ml per tube. Incubate all cultures at 37°C for 6 hr or that length of time previously determined to be the eclipse phase of the virus under study.

c. Discard the medium with sterile technique. Wash the cells with fresh Hanks' medium and discard it. Add 1.0 ml of fresh maintenance medium (i.e., with no pyrimidine) to all tubes of the test and control. Reincubate and observe for CPE for 7 days or appropriate time. Determine titers.

I. ACTION OF ACTINOMYCIN D

Actinomycin D is a polypeptide antibiotic which binds with cellular DNA and inhibits all DNA-dependent synthesis of RNA [40]. The fact that actinomycin D does not inhibit growth of small RNA viruses serves to differentiate replication of viral RNA from synthesis of cellular RNA. RNA viruses differ among themselves in their susceptibility to action of actinomycin D (poliovirus and Mengo virus, for example, multiply normally, whereas reovirus is inhibited).

1. Materials

> Virus suspension
> Uninoculated tissue cultures
> Hanks' maintenance medium
> Actinomycin D
> Test tubes, 12 × 72 mm, sterile
> Pipets, 1.0 and 10.0 ml, sterile
> Test tube rack
> Tissue culture racks

2. Method

a. PRELIMINARY DETERMINATION OF ACTINOMYCIN D
CONCENTRATION

(1) Prepare dilutions of actinomycin D in Hanks' medium at 1.0, 0.1, 0.01, 0.005, and 0.001 μg per ml. Remove the growth medium from 30 tubes of tissue culture. Add 1.0 ml of each dilution of actinomycin D to each of 6 tubes of culture. Incubate as usual.

(2) Determine the concentration least detrimental to the cells.

b. TEST

(1) Prepare actinomycin D-containing Hanks' medium at the desired concentration of the antibiotic just determined.

(2) Discard medium from 64 tissue culture tubes. Retain 10 additional tubes of culture as controls. Add 1.0 ml of actinomycin D medium to the 64 tubes (not the controls). Incubate at 37°C.

(3) Twenty-four hours later, inoculate half of the cultures (32 tubes) with tenfold serial dilutions of virus, 0.1 ml per tube, 4 tubes per dilution, without changing the medium. Forty-eight hours after addition of actinomycin medium, titer virus in the remaining 32 tubes. Observe and record CPE in each set of tubes, for 7 days or appropriate time, and compare with controls.

Histochemical
Methods
SEVEN

On occasion you may wish to study the effect of virus infection, induction of inclusion bodies, replication in the nucleus or cytoplasm, etc., details of which cannot be noted by examination of unstained infected monolayer tissue cultures. You may, then, turn to various histochemical methods as an aid in these determinations. Many procedures are available, but three standard ones will be discussed.

Generally, cultures are prepared in tubes containing small cover slips. The monolayer grows over the "slide," which can be removed from the tube later and processed as any slide for histologic examination by light microscopy. A recent innovation is a sterile plastic compartmentalized chamber mounted on a standard 1 × 3 in. slide (see list of suppliers, p. 272). Several cultures can be prepared on each slide, or several dilutions of the same virus can be inoculated on cells contained in each compartment. The slide is removed and the monolayer can be stained in the conventional manner.

A. PREPARATION OF SLIDE CULTURES

1. Materials

Trypsinized cells
Virus suspension

153

Hanks' maintenance medium, sterile
PBS, sterile
Test tubes, or Leighton tubes, sterile, containing 8 × 22 mm
    (No. 2 thickness) cover slips
Pipets, 1.0 and 10.0 ml, sterile
Test tubes, 12 × 72 mm, sterile
Test tube rack
Tissue culture rack

2. Method

Monolayers of tissue cells should be prepared in test tubes in the usual manner, but with one important addition. An 8 × 22 mm (No. 2 thickness) cover slip should be placed in the bottom of the test tube at the time of assembly for sterilization (Fig. 7-1). Care must be exercised when the tissue culture suspension is initially added and also when the medium is changed to assure that the slide remains in the same position. The monolayer will spread over the surface of the cover slip. You do not want it to be dislodged by having the slide turn over accidentally. This problem may be lessened considerably by the use of special Leighton tubes which have a flattened area designed to contain the slide (Fig. 7-2). These are more expensive and require special racks to hold them. With reasonable care and a little experience, you will find a standard test tube to be adequate.

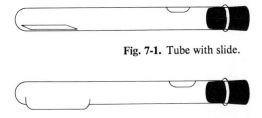

Fig. 7-1. Tube with slide.

Fig. 7-2. Leighton tube.

a. Prepare monolayer tissue cultures in tubes containing 8 × 22 mm cover slips. Incubate until the cell sheet has formed. With care, change to maintenance medium.
b. Inoculate virus as usual. Incubate. Observe.
c. At desired intervals, pour medium into a contamination pan. With care, add PBS to the tube; discard the PBS. Repeat this washing of the cell sheet once more. Hold the slides in PBS until ready for fixation.

B.   HEMATOXYLIN AND EOSIN STAINING PROCEDURE

This stain combination is useful for a general light microscopic examination of cell cultures. Nuclear and cytoplasmic alterations, especially inclusions, are easily seen, and photography of control and inoculated cultures is facilitated.

1. Materials

       Infected tissue cultures, above
       Non-infected tissue culture controls
       Neutral buffered formalin
       Distilled water
       Alum-hematoxylin, 1:30 in distilled water
       Sodium bicarbonate ($NaHCO_3$), 1.0% aqueous
       Eosin, 0.5% aqueous
       Graded series of ethyl alcohol (50, 70, 80, 95%)
       Ethanol:xylol, 2:1
       Ethanol:xylol, 1:2
       Xylol
       Mounting medium (Permount)
       Spatula, $\frac{1}{4}$ in. wide, 6 to 8 in. long
       Forceps, flat-tipped
       Micro-carry-all (Chapter 9)
       Lint-free cloth

2. Method

a. Pour off the PBS. Add neutral buffered formalin to the tube and fix the cells for at least 4 hr at room temperature or overnight at 4°C.

b. Pour off the formalin. Wash the cell sheets for 5 min in three changes of distilled water by gently adding the water directly to the test tube.

c. Remove the cover slip (slide) from the tube as follows:

    (1) Pour off the water.

    (2) Use a long, narrow, flat spatula. Run it down along the glass wall and underneath the cover slip (Fig. 7-3). Try to

Fig. 7-3. Removing slide from tube.

place the flat surface of the spatula directly under the central portion of the slide so that it literally does a balancing act on the spatula. Slowly draw the cover slip slide out of the test tube (Fig. 7-4).

Fig. 7-4. Slide removed from tube.

(3) Carefully grasp the slide on a corner edge using flat-tipped forceps. Place the slide, tissue-side facing you, in a micro-carrying rack (Chapter 9) which is held in a staining dish of distilled water.

(4) Arrange a set of staining dishes according to the protocol in Fig. 7-5, and process the slides as directed.

(5) Place a drop of mounting medium in the center of a 1 × 3 in. microscope slide. Lower the cover slip, tissue-side down, onto the medium. Allow one edge of the cover slip to

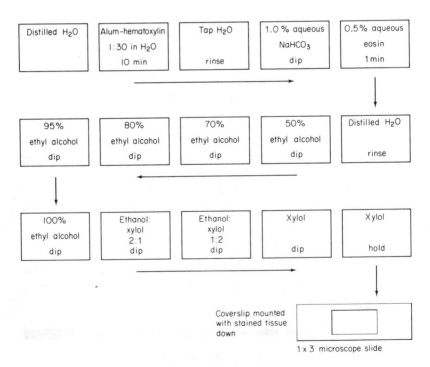

Fig. 7-5. H & E staining procedure.

touch the medium first; then lower the whole slide. This
will help to avoid air bubbles around the tissue (Fig. 7-6).

(6) With a piece of lint-free cloth, grasp the slide and corner
of the cover slip simultaneously in the left hand. Be
firm, otherwise the cover slip will move and cause disruption of the cell sheet.

(7) Dip a far corner of the cloth in xylol and, holding the cloth
over the forefinger of the right hand, stroke over the surface
of the cover slip and slide. This will remove excess mounting medium from the slide and will result in a sparkling
mounted specimen (Fig. 7-7).

(8) Place a permanent label on the left side of the slide. Indicate
the tissue, virus, hours postinoculation, stain, and date.

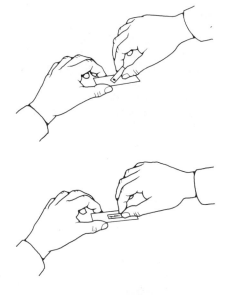

Fig. 7-6. Putting on coverslip.

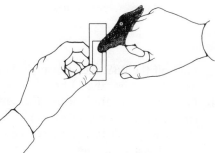

Fig. 7-7. Cleaning the slide.

C. FEULGEN STAINING PROCEDURE [9, 18]

This method has been useful in determining the prime classification of a virus, i.e., whether its nucleic acid is of the RNA or DNA type. DNA viruses multiplying in the nucleus and/or cytoplasm will cause an increased intensity of pink to reddish-purple stained aggregates. It is absolutely essential that the enzymatic tests listed under controls are performed at the same time. Additional uninfected cell cultures and known DNA and RNA virus-infected cultures must likewise be included for comparison. Viruses which are not of the DNA type will not cause the response described above and the cells will remain blue in color. Some viruses, then, can be suspicioned to be of the RNA type and further tests with acridine orange will confirm this test as well as identify RNA viruses.

1. Materials

> Infected tissue cultures, above
> Uninfected tissue culture controls
> Known RNA and DNA virus-infected tissue cultures
> Carnoy's fixative
> Absolute ethyl alcohol
> Graded series of ethyl alcohol (25, 50, 75%)
> HCl, normal
> Schiff's reagent
> Sodium bisulfite, 0.2 M
> Fast green, 0.01% in 95% alcohol
> Absolute alcohol:xylol, 1:1
> Pepsin, 0.02%, in 0.02 M HCl
> Deoxyribonuclease, 0.01%, in 0.025 M veronal buffer plus
>   0.003 M MgSO$_4$
> Ribonuclease, 0.05%, in distilled water, pH 7
> Mounting medium
> Spatula
> Micro-carry-all
> 1 × 3 in. microscope slides

2. Method

a. Pour off the PBS. Add Carnoy's fixative to the tube and fix the tissue for no less than 1 hr.
b. Pour off the fixative. Rinse the cover slip by adding absolute ethyl alcohol to the tube; leave for 3 to 5 min, and pour off.
c. Remove the slide from the test tube as described under hematoxylin and eosin staining above. Place the cover slip slide in the

carrying rack held in a staining dish containing absolute ethyl alcohol.

d. Set up a series of staining dishes according to the protocol in Fig. 7-8, and process the slides as indicated. If the fast green stain is too intense, it can be removed by a water wash, but it will not be removed with an alcohol wash.

e. Controls:

(1) Process matching cover slips with uninfected and RNA and DNA virus-infected cultures simultaneously with the series indicated above.

(2) Process two slides each of infected and uninfected cultures through pepsin plus enzyme as indicated in Fig. 7-9 [34].

(3) Process two slides each of infected and uninfected cultures through enzyme, without pepsin pretreatment, as indicated in Fig. 7-9.

f. Mount all cover slips by inverting them onto mounting medium on a 1 × 3 in. standard slide.

g. Attach a label to the left side and indicate tissue, virus, stain, etc.

D. ACRIDINE ORANGE FLUOROCHROME PROCEDURE

This, like the Feulgen procedure, is used in identifying the nucleic acid type of a virus in infected cell cultures [33]. Where a number of cell cultures are infected and removed from incubation at specified times, an estimate of the rate of viral multiplication can be made by the degree of intensity of staining of either nucleus or cytoplasm. Acridine orange (AO) can also be used in enzyme tests to determine whether the nucleic acid is single or double stranded [34].

Carnoy's fixative appears to be the one of choice; fluorescence is more intense after its use. The buffer used is equally important since prolonged contact with PBS tends to extract AO. McIlvaine's buffer is best.

In uninfected control preparations, the cytoplasm will stain a dull brick red, the nucleus a yellow green, and the nucleolus a yellow orange. In DNA virus-infected cell cultures, compared to uninfected controls, the nucleoli will have increased fluorescence with an intensification of staining to become reddish orange. Infected cell nuclei will show gradually increasing fine accumulations of yellow fluorescing material sometimes filling the entire nucleus. Sometimes inclusions formed will have a bizarre form: half-moon or sickle shape. DNA virus-infected cells appear yellowish green.

Following infection with an RNA virus, increasing cytoplasmic RNA staining will occur with clearing of nuclear chromatin. The cells will begin to

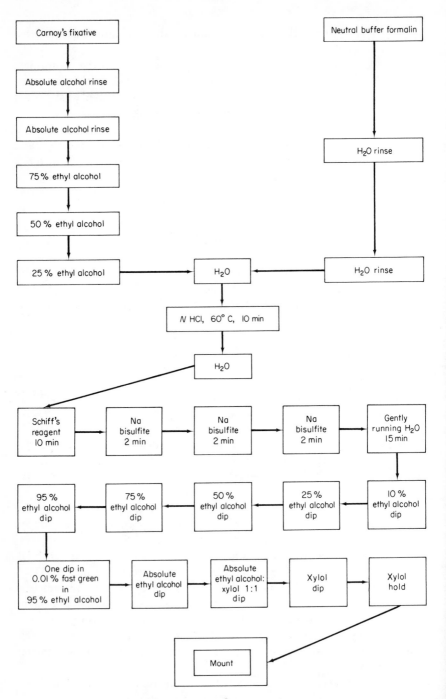

Fig. 7-8. Feulgen staining procedure.

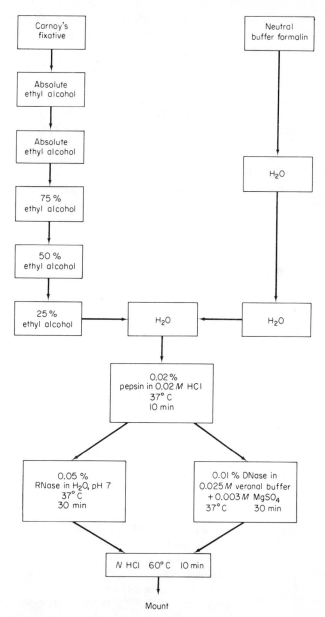

Fig. 7-9. Enzyme control for above procedure.

shrink and the nuclei will become eccentric and brilliant yellow while the cytoplasm will become intensely red. Cytoplasmic inclusions may be noted. RNA virus-infected cells stain orange red. Release of RNA material may be evident as blebs at the cell walls. Following viral release, nuclear material will appear eccentric, pyknotic, and intensely yellow; the cytoplasm may be shrunken, with long cytoplasmic strands, and stain lightly for RNA. Some degree of caution is advised. These reactions may not be evident with all RNA viruses. Cells infected with double-stranded RNA viruses (reovirus, for example) will stain yellowish green. Cells infected with single-stranded DNA viruses (e.g., phage) will stain red. It is advisable, therefore, to use controls with and without pepsin followed by enzyme treatment (Table 7-1).

TABLE 7-1

Staining Properties of Viruses and Cells with Acridine Orange

Nucleic acid		AO	DNase	RNase	Pepsin pretreatment required
DNA	Single-stranded	Flame red	Effect destroyed	No effect	No
	Double-stranded (bacteriophage)	Yellow-green	"	"	"
	Double-stranded (animal virus)	Green	"	"	" (indicates maturation not comlete)
		Yellow-green	"	"	Yes (indicates maturation is complete)
	Host cell	Yellow-green	"	"	No
RNA	Animal virus and/or Host cell	Flame red	No effect	Effect destroyed	No
		(Detection of RNA virus against cell RNA depends on density of staining and distribution of nucleic acid.)			

1. Materials

Infected and non-infected tissue culture cells
Carnoy's fixative
Graded series of ethyl alcohol
Distilled water
McIlvaine's buffer, pH 3.8
Acridine orange, 0.01%, in buffer, freshly prepared before use
Pepsin, as above
DNase and RNase, as above
Mounting medium
1 × 3 in. microscope slides
Spatula
Forceps, flat-tipped
Micro-carry-all

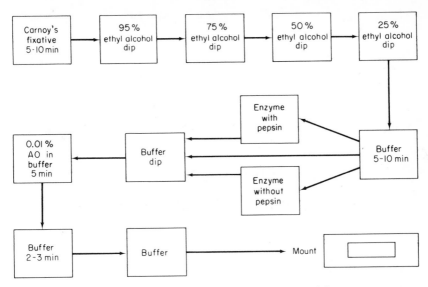

Fig. 7-10. Procedure for acridine orange staining.

2. Method

a. Pour off the PBS. Fix the tissues, without predrying, in Carnoy's fixative for 5 to 10 min at room temperature.

b. Remove the slides to a carrying rack held in a dish of 95% ethyl alcohol.

c. Set up a series of staining dishes according to the protocol in Fig. 7-10, and process the slides as directed.

d. Process uninfected and known RNA and DNA virus-infected tissue culture cells at the same time.

e. Process similar cells through RNase and DNase with and without pepsin as indicated in the protocol.

f. Mount the cover slips by inverting them over a drop of McIlvaine's buffer on a 1 × 3 in. microscope slide. Seal the edges with vaseline, nail polish, or other sealant. Be careful that you do not accidentally move the mounted cover slip as this will smear the tissue. Maintain the slides in a covered Petri dish containing a damp piece of filter paper in the bottom. Observe as quickly as possible to avoid drying and fading of the acridine orange. If delayed, place the slides in a dark place at 4°C until they can be examined.

g. Examine under a fluorescence microscope filtered to emit approximately 4500 Å (angstroms). For color photography, Kodak High Speed Ektachrome 35 mm film may be used. Exposure will vary (the authors have had good results with exposures from 1 to 2 min).

Bacteriophage

EIGHT

In vitro cultivation of animal or plant cells is often expensive, time-consuming, and frustrating. Many find that work with bacterial viruses, or *bacteriophage* ("bacteria-devouring"), is enjoyable and rewarding because *phage* can be grown easily using bacteriologic techniques and simple methods. Fundamentally, both systems are parallel.

The natural environment is an excellent source of bacteriophage. Soil is a good source of phage which infect spore-forming or other soil bacteria; swabs of areas of staphylococcal infection may be a source for phage which infect the staphylococcus bacteria; raw sewage is an excellent source of phage for enteric organisms.

The host bacterium most commonly used for virologic work is *Escherichia coli*, a Gram-negative, non-sporeforming, motile, usually harmless rod. The organism has the advantage of growing rapidly in a simple 1.0% peptone broth solution or in Hanks' medium, cell division occurring every 20 to 30 min.

Several types of *E. coli* phages (called *coliphages*) are known and are designated $T_1, T_2, T_3, \ldots, T_7$ (T for type). The T-phages are tadpole or sperm-like in form, with polyhedral head (about 50 mμ in diameter) and a narrow tail whose length is two to six times the diameter of the head. The head contains a single linear molecule of nucleic acid arranged in a single circle which, in most phages, is double-stranded DNA. The phage tail is literally a hollow tube of protein enclosed within a retractable sheath. *E. coli* B, ATCC (American Type Culture Collection) 11303 and phages T_1 to T_7

(sensitive to all strains) are commonly used, although we suggest the use of T$_2$. Other phage-bacteria combinations are satisfactory: *Bacillus subtilis* (ATCC 15563) and phage 15563 B or *Staphylococcus aureus* (ATCC 23360) and phage 23360 B (d'Herelle's S-6) are good combinations.

Presuming you are not using a commercially prepared phage sample, it is necessary to recover a phage from an environmental source. Under natural conditions, the numbers of phage are not very high so the first step is one of enrichment. The source material (containing some phage) must be incubated with a culture of specific host cells. The phage enter the host and multiply; the host cell ruptures; and new, increased numbers of phage enter the medium. This medium, with more phage in suspension, is centrifuged and filtered to produce a clear phage filtrate. The presence of phage can then be demonstrated in a number of ways, including infection of broth or plate cultures of the host bacterium.

In the phage plaque method, a heavy seeding of host bacteria on an agar surface results in a "lawn" of bacteria. A phage which enters the bacterial cell causes it to burst and 150 to 300 particles released in the "burst" will enter surrounding host cells. These in turn will be lysed and a plaque, or cleared spot, formed by these actively lytic phage can easily be seen against the lawn of bacteria.

Adsorption of phage is influenced by many factors including the composition of the medium, temperature, physiologic condition of the host cells, presence of cofactors, etc. Optimal conditions will, therefore, vary with each phage. Sodium is often necessary in media for phage adsorption, and the presence of tryptone leads to rapid increases in the number of some phage particles. Incubation temperatures vary with the bacterium used although 37°C is standard. *Staphylococcus* phage is grown at 30°C. "Soft" agar (0.75%) is used because more rapid diffusion of the phage can occur. This agar is too soft to use by itself in a plate so it is poured over a base layer of normal, hard (1.5%) agar.

A. ISOLATION

1. Materials

Raw sewage or a concentrated suspension of flies, feces, soil, diseased tissue, macerated cockroaches, etc.

Young (early log phase) culture of *E. coli*, Type B, or appropriate test host

Deca-strength nutrient broth

Membrane filters, two, sterile, porosity determined by host's size

Test tubes, small, sterile

Pipets, 1.0, 5.0, and 10.0 ml, sterile

2. Method

a. Add approximately 45 ml of test material to 5.0 ml of deca-strength broth (this will provide proper concentration of nutrients upon dilution). Add 5.0 ml of early log phase *E. coli* B or other host bacteria, and incubate at 37°C for 24 hr. This is the *enrichment step*. Centrifuge the material at 1400 g approximately 10 min.

b. Set up a sterile protein-treated membrane filter (Chapter 6, Section C) with suction apparatus. Pass the centrifuged material through the filter with water suction. This step is required to remove any remaining host bacteria. Chloroform could be used but is detrimental to some phage. *Save the filtrate!*

c. Add 1 or 2 ml of this bacteria-free filtrate to 10 ml of a young broth culture of host bacteria. Maintain another broth culture of host bacteria as a control, or add a drop of filtrate to an agar plate seeded with a heavy layer of host organisms. This is the *test for the presence of phage* in your sample. Incubate at 37°C, observing every 2 hr. The broth culture may first become turbid, then less turbid, and sometimes clear because of lysis of the bacterial cells by the phage. Clearing or distinct plaques may be seen on agar plates.

d. In the meantime, place the remaining filtrate into labeled tubes and refrigerate. This is a reserve supply.

e. Filter the cleared (lysed) phage-bacterium culture through another sterile filter. You now have a bacteria-free filtrate of phage. Place this in small sterile test tubes and label. The phage may be stored at ordinary refrigerator temperature. You may store stock phage at room temperature in soft agar stabs prepared in 0.25-dram shell vials. Dip a sterile applicator stick or toothpick into the filtrate and use to stab several agar stabs. Seal the vials with tape. Plaques on agar plates may be harvested by stabbing the center of a plaque with a sterile needle, applicator stick, or toothpick. Rinse in broth. This constitutes your *phage stock*.

B. PREPARATION OF INDICATOR BACTERIA

It is advantageous to prepare a stock supply of fresh host cells, particularly if you plan to do a series of tests over a 2- or 3-day period. An important point, usually overlooked when using enteric organisms (and some others as well) is that constant aeration of the cultures must be provided. A simple method is to insert a sterile Pasteur pipet through a one-hole black rubber stopper of a size to fit the culture tubes. Connect this, via sterile rubber tubing, to an air line or a small aquarium pump. Be sure to place a sterile cotton plug or air filter in the line (Fig. 8-1) to prevent bacterial contamination.

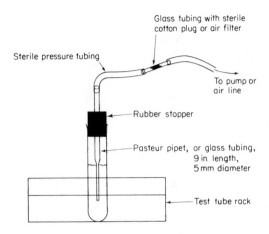

Fig. 8-1. Bubbler tube.

1. Materials

> *E. coli* B or other appropriate host bacterium
> 5 ml nutrient broth, sterile
> 25 ml nutrient broth, sterile
> 250 ml nutrient broth, sterile
> Aeration setup
> Bubbler tubes, sterile
> Centrifuge tubes, sterile
> Pipets, 5.0 ml, sterile
> Water bath, 37°C

2. Method

a. Inoculate 5.0 ml of nutrient broth with *E. coli* B or other host bacterium. Fit the broth tube with a sterile bubbler unit. Place the tube in a 37°C water bath and aerate overnight.

b. In the morning, remove 2.5 ml of overnight culture and inoculate the 250 ml of broth. Aerate for approximately 2.5 to 3.0 hr at 37°C. Pour the culture into sterile centrifuge tubes, cover tightly with aluminum foil caps, and centrifuge at 2600 g for 15 to 20 min. Discard the supernatant material. Resuspend the pellet in 25 ml of sterile broth. These *indicator cells* may be stored in a refrigerator for as long as 3 days.

C. DETERMINING TITER OF UNKNOWN PHAGE STOCK

1. Materials

> Log phase culture of *E. coli* B or other host bacterium
> Phage filtrate (previously isolated)
> Nutrient broth
> Bottom agar, 1.5%
> Top agar, 0.75%, 3 ml per tube, melted, maintained at 47°C
> Test tubes, sterile
> Petri dishes, sterile
> Pipets, 1.0, 5.0 and 10.0 ml, sterile
> Water bath, 47°C

2. Method

a. Prepare three Petri dishes with bottom agar and dry overnight at room temperature or in an incubator at 37°C (if glass dishes are used) or at 45°C (plastic dishes). If agar is not dried before use, plaques will run together, making it impossible to count accurately.

b. Prepare a log phase culture of *E. coli* B or appropriate host cells.

c. Prepare tenfold serial dilutions to 10^{-8} of phage filtrate in broth.

d. Pipet 3.0 ml of hot, melted top agar into three 13×100 mm test tubes held in a 47°C water bath. (As you gain experience, this volume may be cut to 1.5 ml.) Wait several minutes until agar comes to 47°C.

e. To each tube of 47°C top agar, add 0.1 ml of fresh *E. coli* B or other host cells. Always use young indicator cells. Plaque counts decrease if old broth cultures are used.

f. Quickly pipet 0.1 ml of the 10^{-8} phage dilution into a tube of top agar containing host bacteria. Do not put the pipet into the agar! Blow the phage into the tube! Mix well, but gently (Fig. 8-2), to avoid bubbles, and quickly pour the mixture onto the surface of bottom agar (previously prepared) while turning the dish with the left hand. Tap out the final drop, or touch the tube gently to the surface of the agar. Rapidly tilt the plate to distribute the agar over the surface of the bottom layer. Repeat with the 10^{-7} and 10^{-6} phage dilutions (Fig. 8-3).

g. Allow the soft agar to solidify and become firmly attached to the hard agar base. Invert all the plates. Condensation often

Fig. 8-2. Mixing agar in tube.

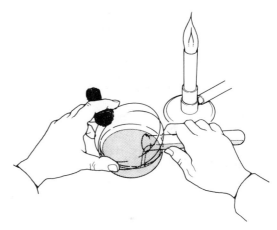

Fig. 8-3. Pouring agar into Petri dish.

develops on the covers as the warm agar solidifies. This condensation may drop onto the surface causing the plaques to run together if the plates are not inverted. Incubate at 37°C, 3 to 24 hr, or until plaques are visible. It is wise to incubate duplicate plates at 40°C since some plaques may not be visible at the lower temperature. Plaque formation can be slowed by placing the plates at a lower temperature when cultures must be left overnight.

h. Determine the titer of the phage:

(1) Look for areas of clearing, or holes, in the opaque lawn of host bacteria. Choose the plate which shows approximately 100 to 200 plaques (large plaque formers) or 300 to 1000 (small plaque formers).

(2) Count the number of plaques seen. Multiply by the final dilution and then times the plating dilution (*total dilution factor*).

EXAMPLE

Plaque count $= 170 = 1.7 \times 10^2$

Dilution containing 170 plaques $= 10^6$

Plating dilution (0.1 ml used) $= 10^1$

Titer of phage stock $= (1.7 \times 10^2) \times 10^6 \times 10^1$

$\qquad = 1.7 \times 10^9$ PFU per milliliter

D. PREPARATION OF HIGH-TITER PHAGE STOCKS

In order to do phage work, it is necessary to use high-titer stocks prepared from single clones. Many variables will determine maximum yield so you will have to develop procedures to fit the phage-host system. You may use either the plate method (reserved for volumes up to 50 ml) or the broth method (for larger yields) to prepare stock phage.

1. Plate Method

The plate method is essentially the same as that used for plaque assay except you will be starting with a phage stock of known titer. You must dilute this to a concentration which will give overlapping plaques. This dilution will vary depending on the phage.

Unlike the plaque assay, the bottom agar should not be too dry because you will be unable to obtain overlapping plaques. The exact stage of dryness can only be determined through experience.

The addition of antiserum (specific for the host cells you use) to the initial phage lysate can be an important step with some systems and should not be overlooked. Bacterial cells, flagella (if present), and debris will separate out much more quickly and easily if antiserum is used.

The final stock suspension must be freed of remaining viable bacteria. The suspension may be filtered; chloroform (1 part plus 19 parts of phage) may be added (but some phage are susceptible to it); or the stock may be placed at 60°C for 1 hr.

A.   MATERIALS

E. coli B or other host bacterium
Phage stock
Antiserum specific for host bacterium
Nutrient broth
Top agar, melted, maintained at 47°C

Bottom agar
Pipets, 1.0, 5.0, and 10.0 ml, sterile
Petri dishes, sterile
Centrifuge tubes, sterile
Chloroform, sterile membrane filter, or water bath at 60°C
Water baths, 37°C and 47°C
Bubbler tubes, sterile

B. METHOD

(1) Prepare a log phase culture of *E. coli* B (*ca.* 5×10^7 or 5×10^8) or other host cells by overnight aeration at 37°C as previously described.

(2) Dilute phage stock to a concentration of approximately 50 to 100 phage per 0.1 ml:

EXAMPLE.

Your phage stock titers at 1.7×10^9.
You desire a titer of 10^9.
You must dilute the original 1.7 times.
You desire 5.0 ml of final 10^9 dilution.
Therefore, dilute 2.9 ml (i.e., 5.0 ml/1.7)
of phage stock with 2.1 ml of broth.

(3) Add 0.1 ml of diluted phage to 3.0 ml (or less) of top agar. Add one or two drops of host bacterial suspension and mix. Pour over a plate of bottom agar, solidify, invert, and incubate overnight.

(4) To each plate, add 3.0 ml of broth (or PBS) and incubate at room temperature for 3 to 5 hr. Gently pour off or pipet the fluid from the plate and place in a sterile centrifuge tube. It is not necessary to harvest the agar itself since its presence adds another step to the procedure (an added annoyance).

(5) Dilute specific antiserum (1280 titer) 1 : 1000 by adding it in proper volume to the quantity of phage lysate you have. Allow to stand at room temperature for 30 min. Centrifuge to remove cells, flagella (if present), and debris.

(6) Collect the supernatant fluid. Filter through a sterile membrane filter; or add chloroform (1 part plus 19 parts of lysate), shake, allow to settle, and remove phage from the top; or place lysate at 60°C for 1 hr.

2. Broth Method

In addition to the need for ample aeration of many bacterial species (as mentioned previously), the time of addition of phage is critical. Try to simu-

172   *Bacteriophage*

late the same conditions used in the plate method above, i.e., bacteria *ca.*
5 × 10⁷ and phage between 50 and 100 per 0.1 ml.

A. MATERIALS

> *E. coli* B or other host bacterium
> Phage stock
> Antiserum specific for host cells
> Growth broth, 25 ml
> Top agar, melted, maintained at 47°C, 3.0 ml per tube
> Bottom agar
> Dilution tubes, sterile
> Petri dishes, sterile
> Pipets, 1.0, 5.0, and 10.0 ml, sterile
> Bubbler tubes, sterile
> Screw-capped tubes, sterile
> Chloroform, sterile membrane filter, or water bath at 60°C
> Water baths, 37°C and 47°C

B. METHOD

(1) Prepare a log phase culture of *E. coli* B (*ca.* 5 × 10⁷) or other host cells by overnight aeration at 37°C as previously described. Refrigerate.

(2) Dilute phage stock to a concentration of approximately 50 to 100 phage per 0.1 ml as described above.

(3) Time 0: Add 0.1 ml of phage (at 50 to 100 phage per 0.1 ml) to 3.0 ml (or less) of top agar. Add one or two drops of *E. coli* B or other host cells from step 1 above, and mix well. Spread this over a plate of bottom agar, allow to solidify, invert the plates, and incubate at 37°C for 5 hr. This is the *preparation of plaques* step.

(4) Time 2½ hr: Inoculate a bubbler tube containing 25 ml of growth broth with 0.1 ml of *E. coli* B or indicator host cells from step 1 above, and incubate at 37°C with aeration.

(5) Time 5 hr:
   a. Attach a rubber bulb to a sterile Pasteur pipet and, with gentle suction, remove material from the center of a well-isolated plaque which should now be visible on the plates prepared at time 0.
   b. Expel this plaque material into the 2½-hr culture of *E. coli* B or other host cells prepared in step 1. Continue the incubation and aeration of this tube at 37°C. The culture will become more turbid and then will gradually clear as cells become infected with the virus and are lysed. Complete lysis will require 1 to 6 hr or more.

(6) Time 6 to 11 hr: When the culture is clear, add specific antiserum

(1280 titer) in a volume to make a 1: 1000 final dilution. Allow to stand at room temperature for 30 min. Centrifuge. Collect the supernatant fluid. Add chloroform and shake well (unlysed cells will then release their phage and viable bacteria remaining will be killed); or filter through a sterile membrane filter; or place lysate at 60°C for 1 hr. Place aliquots of the filtrate into sterile tubes, label, date, and refrigerate.

E. DETERMINING BURST SIZE

ELLIS AND DELBRÜCK [15] developed the one-step growth experiment, a procedure used to determine burst size, i.e., the number of progeny phage produced per infected bacterial cell at the end of the latent period (time from adsorption of phage to release of progeny). This procedure is basic to most studies of viral genetics and should be learned and understood.

Essentially, phage and sensitive bacteria are mixed under conditions which facilitate rapid host cell adsorption of phage, diluted to prevent further adsorption, incubated to ensure lysis of infected cells, and measured to determine total yield of progeny phage. This can be expressed by plotting the count versus time. From the curve, you can calculate the latent period and the average burst size of the infection.

1. Materials

E. coli B or other host bacterium
Phage stock, titer of 2 to 4 $\times$ 10^8/ml, specific for host used
Top agar, melted, maintained at 47°C
Dilution tubes, 9.9 ml broth per tube, sterile, prewarmed to 37°C
Petri dishes prepared with bottom agar
Centrifuge tubes, sterile
Screw-capped tubes, sterile
Pipets, 1.0 and 10.0 ml, sterile
Bubbler tubes, sterile
Water baths, 37°C and 47°C

2. Method

a. Prepare a log phase culture of host bacteria as described previously. Dilute the culture one hundredfold with broth. Incubate the diluted culture at 37°C with aeration for 1.5 to 2 hr. (The culture will become slightly turbid and cells should titer around 10^8 per milliliter.)

b. Prepare a phage suspension at 2 $\times$ 10^8 per milliliter.

c. Have all materials, plates with bottom agar, broth, top agar, etc., prepared and ready to go. Once the following steps begin, you will have no time to stop. Anticipate the number of tubes and plates required and have them ready and numbered. The timing of this and the following steps is critical!

d. Time = 0:
(1) Mix 0.9 ml of diluted *E. coli* B or other host cells with 0.1 ml of high-titer phage stock at 2×10^8 per milliliter or any combination which gives a ratio of host cells to virus of about 10:1. This ratio is called the *multiplicity of infection*, or m.o.i. Too low a ratio will result in many uninfected cells; too high a ratio will result in many deaths by "lysis-from-without" [10]. This happens when bacteria adsorb such large numbers of phage that damage to the bacterial surface occurs with massive leakage of the cell contents.
(2) Incubate for 5 min at 37°C with low aeration. During this time, phage particles are adsorbed by approximately 80% of the bacterial cells (Fig. 8-4).

e. Time = 5 min: Dilute 0.1 ml of the phage-bacterial mixture (above) tenfold (1:100) with broth. This will halt the adsorption process.

f. Time = 5.5 min:
(1) Pipet 0.1 ml of the 1:100 suspension above into 9.9 ml of broth (1:10,000). Mark as Growth Tube No. 1. Aerate at 37°C.
(2) Add a few drops of chloroform to a 2.0-ml sample of this 1:10,000 suspension, and shake the contents. Allow the chloroform to settle. Add 0.1 ml of the supernatant phage plus one drop of indicator host cells to 2.5 ml of top agar held at 47°C. Pour over bottom agar and spread quickly. Allow the agar to solidify. Invert. Incubate at 37°C overnight. Determine the *titer of unadsorbed phage*.

g. Time = 7 min: Pipet 0.1 ml of the 1:10,000 sample from Growth Tube No. 1 above into 9.9 ml of broth (1:1,000,000) and mark this as Growth Tube No. 2. Aerate at 37°C.

h. Time = 10 min:
(1) Pipet two 0.05-ml samples of fluid from the aerated Growth Tube No. 1 into two tubes of 2.5-ml top agar held at 47°C. Add one drop of indicator cells to each. Mix.
(2) Quickly pour onto two plates of bottom agar, allow to solidify, invert, and mark as Plates No. 1a and 1b. Incubate at 37°C overnight.
(3) Determine the *titer of total infective centers*, i.e., infected cells plus free phage.

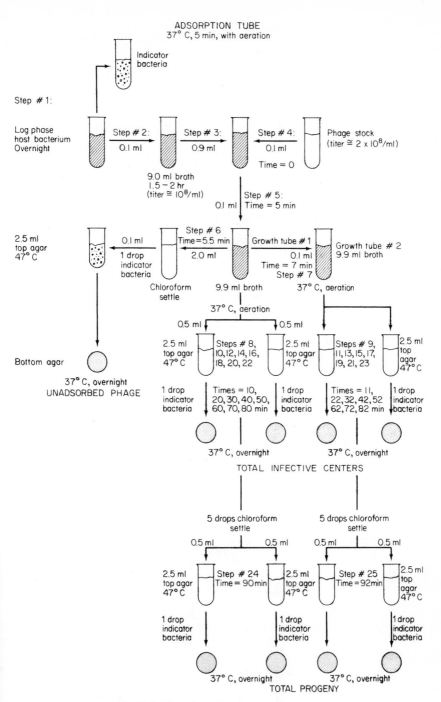

Fig. 8-4. Flow sheet for phage study.

175

EXAMPLE

Titer of total infective centers =
number of plaques $\times$ dilution (10^4) $\times$ plating dilution (0.05) =
number of plaques $\times$ 10^4 $\times$ 20.

 (4) Determine the titer of infected bacteria by subtracting the titer of unadsorbed phage from the total infective centers titer.

EXAMPLE

Unadsorbed phage titer =
number of plaques $\times$ dilution (10^4) $\times$ plating dilution (0.1) =
number of plaques $\times$ 10^4 $\times$ 10.
Therefore, number of plaques $\times$ 10^4 $\times$ 20 $-$ number of plaques $\times$ 10^4 $\times$ 10 = titer of infected bacteria.

 (5) If you wish to measure the length of the latent period, and follow the rise period, you should repeat the titering for infective centers during the next $1\frac{1}{4}$ hr as follows:

Time (min)	Growth Tube No. 1		Growth Tube No. 2	
12			0.1 ml	Plates 2a 2b
20	0.1 ml	Plates 3a 3b		
22			0.1 ml	Plates 4a 4b
30	0.1 ml	Plates 5a 5b		
32			0.1 ml	Plates 6a 6b
40	0.1 ml	Plates 7a 7b		
42			0.1 ml	Plates 8a 8b
50	0.1 ml	Plates 9a 9b		
52			0.1 ml	Plates 10a 10b
60	0.1 ml	Plates 11a 11b		
62			0.1 ml	Plates 12a 12b
70	0.1 ml	Plates 13a 13b		
72			0.1 ml	Plates 14a 14b
80	0.1 ml	Plates 15a 15b		
82			0.1 ml	Plates 16a 16b

i. Time = 90 min:
(1) Add 5 to 10 drops of chloroform to the growth tubes, shake the tubes, and allow the contents to settle. The chloroform will induce lysis of infected cells which have not lysed spontaneously.
(2) Plate 0.1 ml of lysate on indicator lawn, as above. Incubate overnight and calculate the *total progeny* (progeny phage plus unadsorbed phage).
(3) Determine the *burst size*.

$$\text{Burst size} = \frac{\text{progeny phage}}{\text{infected bacteria}}$$

$$= \frac{\text{total progeny unadsorbed phage} - \text{unadsorbed phage}}{\begin{array}{l}\text{total infective centers (infected} \\ \text{bacteria plus unadsorbed inoculum)} \\ - \text{unadsorbed phage}\end{array}}$$

F. PHAGE TYPING

Phage are highly specific for a given strain or species of bacteria. You can make use of this specificity for identification purposes. Epidemiologic studies can be performed using phage-typing methods.

1. Materials

Broth culture of different strains of *E. coli* or unknown
   bacterium to be tested
Phage suspensions (specific types)
Poured plates of top agar on bottom agar base, one for each
   bacterial strain to be tested
Cotton swabs, sterile
Broth, sterile

2. Method

a. Dip a sterile swab into a broth culture of the bacterial strain to be tested and gently, but thoroughly, swab the surface of an agar plate (Fig. 8-5). Invert the plate, tilt the bottom slightly open, and dry at 37°C for 1 hr.
b. With a wax marking pencil, mark the bottom of the dish into squares. Label one square *Control* and the other squares with the specific phage strains being used (Fig. 8-6) or use sterile plastic plates with gridded bottoms.

Fig. 8-5. Swabbing agar plate.

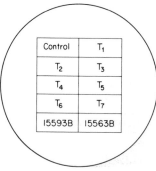

Fig. 8-6. Marking Petri dish.

c. Sterilize an inoculating loop in a flame and, with sterile technique, remove a loopful of phage type and place in its appropriate square. Place a loopful of sterile broth in the control square. Repeat with all phage strains [Fig. 8-7(a), (b)].

d. Allow the inoculated areas to dry at room temperature. Incubate the plates at 37°C for 18 hr. Determine the strain of unknown bacterium.

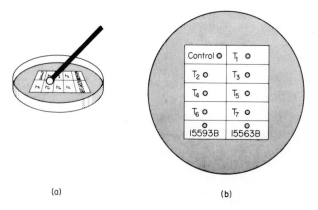

(a)   (b)

Fig. 8-7. Inoculation of phage typing plates.

Plate A, seeded with an unknown strain of
E. coli, is inoculated with phages T_1 through
T_7, phage 15593 B (sensitive to *E. coli*,
ATCC 15593), and phage 15563 B (sensitive
to *B. subtilis*, ATCC 15563). Following incubation,
  Squares T_1 through T_7 show plaque formation.
  Square 15593 B has no plaques.
  Square 15563 B has no plaques.
Therefore, the unknown *E. coli* is strain B
(which is sensitive to all strains of T-phage).

Preparation of Materials, Equipment, and Supplies

NINE

The greatest task facing the researcher is that of "getting ready" to do virus work. Accumulating the tools of the trade, cleaning, assembling, and sterilizing them, and preparing and sterilizing media probably take more time than the actual virologic techniques which follow.

In actuality, this is as it should be because it is on this foundation of scrupulously clean and sterile glassware, proper media, etc., that one must build one's house of virology. Without this proper foundation, all other building attempts will fail. You must not minimize the importance of these foundation procedures. As too often happens, the preparative materials are processed by trained personnel, and come to the worker all "ready to go." The student never learns this most important aspect of the work and never, therefore, appreciates the effort that goes into it, the pitfalls that can arise, or how one can correct them. It is to avoid this knowledge gap that preparation of your foundation materials is given here.

A. PRECAUTIONS

Lack of success in growing tissues in culture can more often be attributed to unclean glassware than to any other problem. The proper cleaning of laboratory glassware can not be overstressed. Contamination with bacteria and toxic substances, such as chemicals, adds to the problem.

Toxicity is often related to the chemical composition of the glass used.

Most tissue culturists prefer to use hard glass (Pyrex, Kimax) rather than soft glass because the latter liberates large amounts of alkali with repeated sterilization. The alkali, soluble in solution, changes the pH of the medium in which the cells grow and metabolize. Hard glass withstands repeated sterilization, contains no free alkali, and has a low coefficient of expansion and greater mechanical strength. Glassware can be expensive in terms of manhours for its preparation (washing, packing, sterilization, etc.) for use.

Plastic ware is now used in many laboratories. There are several varieties, and each has its own advantages and disadvantages. The commonest type used in tissue culture work is probably that made of polystyrene, a rigid, crystal clear, non-toxic plastic used for "disposable" dishes, culture chambers, etc. This plastic cannot withstand high temperatures; when autoclaved to remove contaminating organisms, the plastic melts and can't be reused. For those laboratories where supply monies are readily available, it is suggested that the various plastic wares for tissue culture be used. For those who find monies are more readily available for personnel than for supplies, the use of standard glassware may prove more convenient. Which is more economical in the long run must be an individual decision.

Of equal concern is the water used to wash glassware and prepare media. Tap water must never be used. It often contains microorganisms, dissolved organics (detergents, wastes), dissolved inorganics, and particles of various kinds from pipes and valve connections, etc. In many areas, calcium or magnesium salts in the water supply will deposit on glassware. This leads to clouding of the glass or to loss of the surface in soft-type glass. These carbonates change the media placed in such treated glassware and result in death of tissues in culture. Media, therefore, cannot be considered standard from one city to another or from one country to another.

Particulate materials are found to contaminate effluent of a still. Some organics may be carried over with the vapor phase and inorganic salts from "carry-over" and the still materials themselves (10 to 50 μg per liter copper, for example) may be present. Distillation doesn't normally remove carbon dioxide or ammonia from water. Raw water carbonates (mentioned above) decompose to carbonates in the still boiler. Therefore, glass-distilled water should be passed through ion-exchange resins (charged with two parts per volume of a strongly basic anion exchange resin in the hydroxyl form and one part by volume of a strongly acidic cation exchange resin in the hydrogen form). Use this deionized-distilled water to prepare all solutions and to final rinse all glassware.

The type of detergent used is also an important and often critical consideration because it will adhere to glass surfaces, and most detergents are toxic for cells. There are times when glassware is washed and rinsed with utmost care and cultures still will not grow. When this happens, it is suggested that different types of detergent be tried until one is found that is alkali-free, lowers the surface tension, has good wetting ability, softens the water, and is soluble in both hot and cold water.

It should be common practice to keep tissue culture stock chemicals and media on a separate shelf or even under lock and key. These chemicals should be used by no one else. Far too often, workers cross-contaminate chemicals because they fail to wash spatulas thoroughly, do not recover bottles immediately, etc. Chemicals or media which are hygroscopic should be kept in a desiccator over calcium chloride crystals. Alternatively, the covers should be closed tightly and sealed with tape or dipped quickly into hot, melted paraffin to a depth of at least 1 in.

Keep the pans and the platforms of analytical balances scrupulously clean. Many powders are so fine that the slightest movement may cause particles to fall into the chemical you are weighing. Insist that a record book be kept next to the balance. All users should sign it each time the balance is used, and they should indicate the type of chemical weighed at that time.

It is necessary to use disinfectants at various times on work surfaces and other inert surfaces (contaminated pipets and other glassware, for example). See section G for a full discussion of methods of disinfection.

B.   CLEANING PROCEDURE—NEW
GLASSWARE

1. General

Once new, unused glassware has received the following treatment, the procedure need not be repeated. Rinse new glassware thoroughly in tap water and then completely immerse it in a 4% solution of hydrochloric acid (HCl) in water and leave overnight to soak. A large plastic garbage container with cover is quite satisfactory for this because long graduate cylinders and large Erlenmeyer flasks can be submerged with no difficulty.

The following morning, rinse the glassware thoroughly in running tap water and then submerge it in detergent solution. Large amounts of detergent are not necessary; i.e., you do not need volumes of suds in order to effect proper cleaning. Add only that amount of detergent which will give a solution which is slippery to the fingers—that much, and no more. Leave the glassware submerged in detergent overnight.

The following morning, rinse the glassware thoroughly, inside and out, at least 10 times with tap water, then 5 times in double-distilled water, and once in deionized water. Invert all glassware and place in wire baskets or hang on pegs to dry. It can then be processed for sterilization (see Section F).

2. Pipets, Cover Slips, and Slides

Glass pipets may be similarly treated in a large, deep tray and then rinsed in commercially available pipet washers. *Be careful when you remove the pipets from the detergent solution to the pipet washer. The pipets are very slippery to the touch and will easily fall from your grasp!* Place the pipets in the washer with the mouthpiece down and the tip end up. This permits the water to flush upward through the pipets with ease. Washers with a heating and drying cycle are best. Standard washers may be connected to a hot water line, however, and the wash cycle maintained for at least $\frac{1}{2}$ hr.

In the event that a commercial washer is unavailable, you can use a water suction apparatus (Fig. 9-1). Use a hose connection with an inside diameter small enough to accommodate the tip end of a pipet. It may be necessary to adjust the tubing downward in size by inserting small lengths of glass tubing in the line.

Place the tray of pipets under hot, flowing, tap water or carefully hold individual pipets under flowing tap water, in order to remove detergent from the outer surface of the pipets. Place in a clean tray.

Fill a large beaker with hot tap water, another with distilled water, and

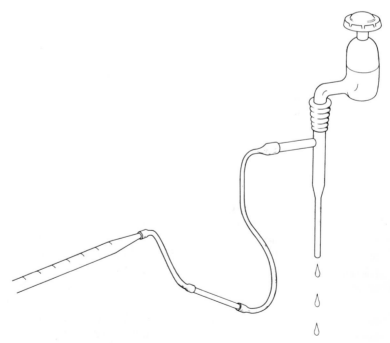

Fig. 9-1. Water suction apparatus.

another with deionized water. Turn on the tap water to obtain adequate suction. Place the pointed end of the pipet into the suction hose. Hold the mouth end of the pipet in the beaker of hot tap water and draw the water through while raising and lowering the mouth end in and out of the water. This will give a more vigorous flushing action. Move the pipet sequentially through the water series held in each beaker using the same type of in-and-out motion. It is suggested that you draw water through each pipet at least 15 times per beaker of water to assure a proper rinse. Finally, draw air through the pipet to remove excess water.

Place the pipets in a pan or wire basket and place in an oven. Turn the temperature to 160°C (325°F in a home-type oven) and leave until the pipets are completely dry.

Acid-soak slides and cover slips of non-corrosive hard glass for 30 min, rinse in tap water, and then boil 10 min in a dilute detergent solution. Then rinse them in distilled or deionized water, immerse in 70% alcohol, and dry by holding each cover slip or slide with forceps and passing it quickly through a Bunsen flame. Alternatively, each may be wiped dry with a lint-free cloth.

C. CLEANING PROCEDURE; CONTAMINATED–USED GLASSWARE

In the case of material which is contaminated by contact with virus, it should be routine procedure to place pipets into a basket held in a pipet jar filled above the level of the pipets with disinfectant or a detergent solution containing 0.5 to 1.0% phenol.

Remove pipets, which have been immersed in disinfectant, to a shallow pan. If a highly pathogenic virus was used, handle the pipets only with gloved hands. Place such pipets into a contamination pan containing detergent solution, slide your gloves off and into the pan as well. Then autoclave the pan and its contents at 15 lb pressure for 15 min.

If you have used plugged pipets, the cotton plugs in the mouthpiece may be removed by inserting the pointed end of the pipet into a rubber hose attached to an air line. A small amount of air pressure is usually sufficient to blow the plug out through the mouthpiece and into a pan or other container. Do not blow plugs into a sink where they will stop up the drain! In the event of an obstinate plug, the easiest device to use is a fine, No. 13 steel crochet hook, which can be obtained at most sewing counters. This is thin enough to go into the mouth of the pipet, yet the hook is strong enough that it will not bend. The cotton can be hooked and gently pulled out with this. Place the pipets in an automatic washer with the mouthpiece down and the pointed (tip) end up. Process as discussed previously.

Place test tubes, rubber stoppers, etc., immediately after use into a contamination pan with a cover. Needles and syringes (with plungers removed) should be placed in disinfectant in a smaller pan inside the contamination pan. Preferably, these pans should be of stainless steel because steel won't chip after continued use (chipped areas make excellent hiding places for bacteria and viruses); the pan has smooth, rounded corners; and it can be ordered in a size which will fit your autoclave. To ensure that materials do not dry on the glassware, and as an aid to cleaning, add detergent solution to the pan. Do this through a corner opening (i.e., with the lid drawn slightly aside) so that splashing of infected material does not occur. Autoclave the pan and its contents at 15 lb pressure for 15 min. Remove the pan and allow it to cool or place the pan in a deep sink and let cold tap water flow onto and over the cover until the contents have cooled sufficiently to be handled. Process the glassware by hand or by an automatic washer as described below.

D. CLEANING PROCEDURE;
NON-CONTAMINATED–USED GLASSWARE

Immediately after use
1. Put all used materials into a pan of detergent and water. Separate syringes so that all surfaces are exposed. Submerge all small pieces of glassware and completely fill with water. Do not leave a large air pocket inside flasks or tubes; otherwise materials will dry and harden on the glassware and "bathtub" rings will occur.
2. Rinse large flasks, graduate cylinders, etc., in running tap water and then fill to the top with detergent and water, or at least, water.
3. Submerge pipets in detergent solution in a pipet basket held in a pipet jar.

Then routine washing should proceed as follows:
4. Wash glassware in hot detergent solution. If hand-washing, use test-tube brushes (*reserved for this purpose only!*) *on* and *in* all test tubes and flasks. In the event of a particularly stubborn area, immerse the culprits in a pan of detergent solution and autoclave at 15 lb pressure for 15 min. *Be sure you do not use a plastic wash pan as it will melt!*

If you have automated glassware washers, follow directions of the manufacturer. The problem with most of these machines is that the water is never hot enough (it should be at least 160°F) when it hits the glassware, and there should be no obstruction to water flow into small tubes, flasks, and cylinders. If the washer is attached to a steam line, you are most fortunate. To be effective, however, the machine should be adjusted so that the water is superheated, i.e., above the temperature of boiling water (212°F, 100°C).

If you are fortunate enough to own a Char-Lab, you have the best washer ever made, in the opinion of the authors. Rapid flushing action in and around glassware occurs and, attached to a steam line, superheated water can be used. Many people have thrown them out in favor of the new automatics which usually can't do as good a job because they can not duplicate the same type of flushing action as the Char-Lab. Unfortunately, this unit is no longer manufactured.*

5. Place pipets, tip end up, in an automatic washer and rinse for at least ½ hr as described previously.
6. Rinse all glassware under copious amounts of hot running tap water.
7. If there is any problem with the water supply, rinse the glassware in three changes of distilled water and one change of deionized water.
8. After washing, thoroughly dry all glassware before it is wrapped for sterilization purposes. Turn test tubes, flasks, etc., upside down on trays or place them in wire baskets or on pegs to drain. All this material can be oven-dried if you place it in a cold oven and gradually turn up the temperature. Air-dry any materials which would be ruined by heat (coated stirring bars, rubber tubing, etc.).

E.   MISCELLANEOUS EQUIPMENT

There are various smaller items required in the laboratory which, while they require no special handling, should be discussed. These are items which are small or made of materials other than glass.

1. Stoppers—Rubber

Black rubber stoppers are usually used for flasks or bottles containing media or in those situations where the stoppered article does not come in contact with cells or virus. These stoppers are not made of pure rubber and do not withstand repeated autoclaving. They may also have a high content of sulfur or other components toxic for cells.

White rubber stoppers are normally used for tissue culture tubes or bottles or in those situations where the stoppered article comes in contact with cells or virus. These stoppers are more expensive because they are of pure gum, are non-toxic to cells, and withstand repeated autoclaving.

Wash black or white rubber stoppers separately in hot running tap water, rinse in distilled water, and dry on paper toweling. If they are particularly soiled, detergent may be employed. Do not permit rubber stoppers to soak in detergent, however, since detergent becomes absorbed by the stopper and may be a cause of culturing problems later on.

*Formerly available from Fisher Scientific Co., 461 Riverside Ave., Medford, Mass.

2. Plastic Caps

Black plastic bottle caps normally require minimum cleaning because they do not come in contact with adhering materials. When necessary, however, rinse them quickly in running water and turn over on paper toweling to drain. If the liner becomes dry or shriveled, discard the cap. Replacement liners are available but the authors have not found them to be worth the time and effort required to glue them into place.

3. Syringes

Separate standard glass syringes and wash with detergent in the usual manner. If the user has failed to separate the syringe, and the barrel is caked with dried blood or other material, it may be necessary to heat the pieces in a small pan of detergent solution and then use a small bottlebrush to scrub them. Rinse well in tap water and then distilled water and dry separately.

Cornwall syringes may be cleaned in the following way. It is most important that water be pumped through the two-way valve (see Section H), rubber hose and metal ball, by placing the ball in detergent, then tap water, and then distilled water, each time manipulating the syringe and expressing the water into a sink. Disconnect all pieces and allow to dry. Be extremely careful that you do not lose the small springs, rings, and rubber tack valves.

4. Needles

A needle may be cleaned by attaching it to a syringe and aspirating detergent, then tap water, and then distilled water through it. If the needle is plugged, use the special wires supplied with the needles to ream out any material inside. In the event of a badly encrusted needle, it can often be saved by autoclaving it in a pan of detergent solution. Then aspirate water through the needle as discussed above.

5. Plastic Ware

Some plastic ware can be autoclaved. Plastic ware designated "disposable," if contaminated with virus, must be autoclaved and will melt. "Disposable" plastic ware which has not been contaminated can be washed and reused. Such plastic can not be heated or flamed for sterilization purposes. These may be sterilized by placing them under an ultraviolet light (see Section F).

6. Rubber Gloves

Rubber gloves may be washed, while they are still on the hands, with hot detergent and water and then rinsed. Pat dry with paper toweling. Then remove the gloves by grasping them at the cuff. Pull the glove down and over the fingers. The glove should then be inside out but some of the fingers will probably still be inverted. Grasp the cuff of the glove in both hands, swing the glove in a circle away from you, thereby trapping air within the glove. Close the cuff area tightly in one hand and use the other hand to apply pressure to the area just below the fingers. This will force the air into the fingers and they will pop outward. Sprinkle talcum powder into the glove and then outside the glove. Wrap and sterilize (see Section F).

Gloves which have been used to handle contaminated materials should not be washed while still on the hands. Remove the gloves by grasping them inside the cuff. Carefully pull the glove down and over the fingers. Let the glove fall gently into a container of disinfectant and then autoclave them (see Section F). You can then treat them as indicated above.

7. Rubber Tubing

Rubber tubing, like rubber stoppers, should not be placed in detergent solution for any period of time. Rinse the tubing under rapidly flowing water, letting the water flow through the bore. Hang the tubing up to drain and permit it to air-dry. For sterilization, see Section F.

8. Instruments

Forceps, scissors, spatulas, etc., often become coated with pieces of tissue which do not flush away under flowing water. A piece of fine steel wool may be used to remove this. The authors find that commercially available abrasive powders will help to remove the tissue and will polish the instrument at the same time. Rinse the instruments in tap water and wipe dry with paper toweling. Instruments such as these should be cleaned immediately after use otherwise the tissue dries, becomes difficult to remove, and the instruments become rusted. If the instruments have been in contact with infectious material and flaming in alcohol is not feasible, place them in a pan of detergent solution and autoclave according to Section F. Then wash them in the usual manner.

F.   STERILIZATION

Sterilization is the method by which you free any object or substance from all life of any kind. It is important that you have knowledge of methods of killing, removing, and/or inhibiting the growth of microorganisms. You must prevent contamination of all materials used in pure-culture work. A variety of materials (glassware, filters, media, serum, tissues) will be used, each of which is different chemically and, therefore, requires different sterilization processing. Remember: just one organism and an object is no longer sterile!

Heat is probably the most common method of sterilization (but in special cases chemicals, filtration, or irradiation may be used). Heat may be applied in three ways: (1) moist heat (steam or hot water), (2) dry heat (oven), and (3) incineration (flame).

1. Heat—Moist

The method of most concern to you is one which involves the use of compressed steam. You are probably familiar with its use in a home pressure cooker which is a simple form of autoclave. And, indeed, where funds and/or space are severely limited, a large home pressure cooker will suffice.

Both the pressure cooker and the large laboratory apparatus are designed to heat articles using steam under considerable pressure. Steam under pressure is hotter than boiling water (which can never reach a temperature above 100°C or 212°F). The higher the steam pressure, the higher the temperature you can achieve (Table 9-1).

It is important for you to realize that compressed steam (moist heat) sterilizes and compressed air (dry heat) does not. Pure steam at 15 lb pressure has a temperature of 121°C (Table 9-1). If this steam is mixed with 50% air *at the same pressure*, the temperature will drop to 112°C; if there is no air discharge, the temperature will be 100°C. You must leave the exhaust valve open until all air has been driven out of an autoclave (or cooker); otherwise the pressure will not rise and the proper temperature will not be reached. With either piece of equipment, once the pressure is up, the exhaust valve should be closed and you should always observe a fine flow of steam from the

TABLE 9-1

Effect of Air on Temperature At Various Pressures

Gauge pressure (*lb*)	Pure steam, complete air discharge		One-half air discharge 15-in. vacuum		No air discharge	
	°C	°F	°C	°F	°C	°F
5	109	228	94	202	72	162
10	115	240	105	220	90	193
15	121	250	112	234	100	212
20	126	259	118	245	109	228
25	130	267	124	254	115	240
30	135	275	128	263	121	250

bypass valve or from the safety valve; otherwise the autoclave or pressure cooker will explode! It is essential that both be supplied with a thermometer which is more important, from a sterility point of view, than the pressure gauge. The pressure gauge and the safety exhaust valve are important to the safety of operation.

The autoclave is usually operated at 15 lb pressure (1 atm), 121°C, for 15 to 20 min but may range from 10 lb (115°C) to 20 lb (126°C). The following important precautions should be noted:

a. Arrange packages, flasks, etc., so that steam can circulate freely. Articles which are tightly packed or are in closed containers will not be sterilized at the end of normal operation. They will require higher pressure and/or longer sterilization time.

b. Fill flasks or other containers of liquids no more than two-thirds full; otherwise the liquid will overflow the container during autoclaving.

c. Liquids processed in small volumes will require normal time and temperature for sterilization. Large containers with large volumes of liquid will require higher temperature and a longer time to effect sterilization.

d. If you are processing both wet and dry materials through the autoclave simultaneously, always process at the temperature and time to sterilize the wet materials.

e. Completely evacuate air from the chamber.

f. A fine flow of steam should always be seen flowing from the bypass valve.

g. Begin timing only after the *temperature* reaches 121°C (at 15 lb pressure).

h. At the end of the sterilization time, allow the pressure to drop slowly. This is particularly important when liquids are processed. Superheated fluids will boil over if the pressure drops rapidly and much of the material will be lost in the autoclave; cotton plugs will become wetted and/or will blow out of tubes and flasks; glassware may crack.

Operation of an autoclave is not discussed. Many models are available; some are electrically heated, others are plumbed to a steam line, some operate automatically, and others must be controlled manually. You are urged to familiarize yourself with the operation of the particular model at hand.

One point of technique which should be mentioned is one which is often overlooked, namely, proper opening of the autoclave door. Even though the pressure gauge indicates that the chamber is evacuated of steam, this is not entirely true. There is residual steam in the chamber. It is wise to develop the habit of standing to the right and behind the door (or to the right of those doors which roll to the inside of the chamber) as you open it. In this way, any residual steam will flow out into the room rather than into your face or over your hands. A steam burn is most painful!

2. Heat—Dry

Dry heat is the type used in oven sterilization. In the absence of a laboratory-type oven, a kitchen-type stove or wall oven may be used. Since there is no moisture involved, temperatures must reach 165° to 170°C (329° to 338°F) in order to free articles of spores. It is common practice to use 165°C for 2 hr; in a home-type oven, 350°F is satisfactory. At this temperature, paper wrappings, cotton, string, or gauze will turn slightly yellow. Place a thermometer in either type oven as a check on the control dial. Avoid overheating, which results in charring of paper or cotton and release of cotton oils, or underheating, which results in lack of sterilization. Be sure that if you use a home-type oven you *do not place paper-wrapped packages near the heat coils* because *the paper will catch fire!*

It is important to realize that the 2 hr are timed *after the materials reach the correct temperature.* If articles are placed in a cold oven and, indeed, even if they are placed in a hot oven, there *must* be an interval of time during which all surfaces of all articles are reaching maximum temperature. Do not overload the oven. Allow space between items on shelves. Do not open the oven door during sterilization; otherwise the articles will cool and effective sterilization cannot be assured. In practice, therefore, you must actually leave articles in the oven for 3 to 4 hr. The authors use 4 hr with a fully loaded oven. In any case, be sure to place a sign on the oven door which reads "*Sterilizing! Do not open! Time*: 9 : 00 A.M. to 1 : 00 P.M," for example.

All flasks, pipets, test tubes, etc., should be dry before hot-air sterilization to prevent decomposition of the glass and/or breakage. After sterilization, allow the oven to cool somewhat before the door is opened to prevent sudden contraction and splintering of glass. Remember that you cannot place hot glassware on a cold surface without the strong possibility that the glass will crack. Remove the glassware to a surface which is not cold (asbestos pad, toweling), place a warning sign on it that it is *hot and sterile*, and allow it to cool. It is a better procedure to turn the oven off and let the material cool gradually within. A convenient procedure is to place the glassware in an oven with an automatic timer shutoff before leaving for the day and the glassware will be ready for use in the morning. (Once again, remember to leave a sign on the door with the time that the oven will shut off in case the night watchman becomes concerned.)

It should be routine practice to employ sterility checks during heat sterilization. There are commercially available pieces of paper, thread, or gauze with thermolabile dyes which turn color if the contents have been subjected to the correct temperature for the correct length of time. Merely place the indicator strip in the autoclave or oven before closing the door. A more time-consuming method is to use filter paper strips impregnated with spores of *Bacillus subtilis* or sealed ampuls containing spores of *Bacillus stereothermophilus*. Insert these in the material to be sterilized. After sterilization, aseptically remove the strips to tubes of culture media and incubate at 37°C for 7 days or remove the ampuls to a 55°C incubator for the same length of time. If the media or ampuls are free of growth after 7 days, your autoclaving technique has been satisfactory.

Another method is to aseptically remove a sample from a "sterilized" liquid and inoculate an agar slant (for aerobic organisms) and a deep thioglycollate broth (for anaerobic organisms) [Fig. 9-2(a)–(d)]. Incubate at 37°C (55°C for thermophiles) for 7 days. If no growth is observed in either medium, your sterilization process has been satisfactory.

3. Heat—Incineration

Incineration (heating in an open flame) may be used for mouths of flasks and test tubes and to sterilize sheets of aluminum foil, forceps, scissors, etc. Forceps and scissors should first be placed with just the working portion immersed in 95% alcohol. Holding the forceps or scissors in a *downward* position, run them through the flame and allow the flame to burn itself out (Fig. 9-3). Be sure that you *do not* tilt the forceps point upward because the alcohol will flow toward your hand and you may find your hand and possibly the sleeves of your lab coat being burned in flaming alcohol.

Hold inoculating loops, should you require their use, so that the needle is at an approximate 80° angle to the flame. The needle will enter the flame with the point slightly ahead. Allow the entire length of the needle to become glowing red before removing it from the flame (Fig. 9-4).

Hold sheets of aluminum foil, or aluminum foil bottle covers, etc., in the hands and hold the foil to the flame until it becomes hot to handle.

Pass the mouths of sterile flasks and test tubes through a hot flame periodically during use (transferring fluids, adding or removing fluids, etc.). Essentially, the heat kills any attached bacteria which might accidentally drop thereon during handling. The glassware should not be held in the flame, however, particularly if liquids are to flow in or out because the glass will shatter.

When using an open flame for sterilization purposes, be sure that you *do not have ether or other flammable materials in the room*! When using alcohol, be extremely careful to place the container in such a position that you cannot knock it over; otherwise the work area will go up in flames. The authors have found that a short, heavy, wide-mouthed specimen jar, fitted with a cork stopper is adequate to hold alcohol. A small test tube can be attached to the side with masking tape and forceps can be kept in the tube. The bottle can,

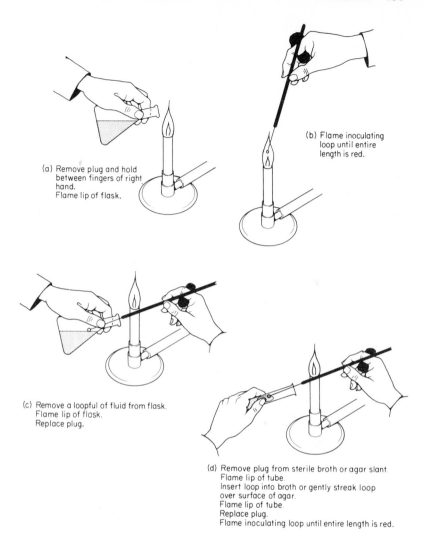

(a) Remove plug and hold between fingers of right hand.
Flame lip of flask.

(b) Flame inoculating loop until entire length is red.

(c) Remove a loopful of fluid from flask.
Flame lip of flask.
Replace plug.

(d) Remove plug from sterile broth or agar slant.
Flame lip of tube.
Insert loop into broth or gently streak loop over surface of agar.
Flame lip of tube.
Replace plug.
Flame inoculating loop until entire length is red.

Fig. 9-2. Method for inoculation.

therefore, be kept closed when not in use, and accidental spillage is kept to a minimum.

The use of a "touch-o-matic" type of burner is preferable. This can be adjusted for a constant flame or, if constant flame is not required, a pilot lamp can be kept lit. A wide touch-plate is positioned at a convenient distance from the flame and you can hold this down with the edge of the hand while simultaneously holding and flaming a flask or pipet held in the right hand.

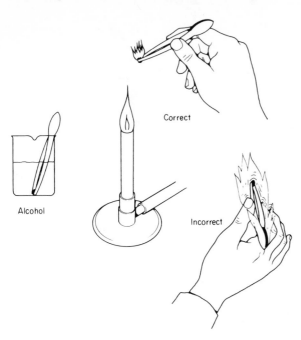

Fig. 9-3. Correct flaming of forceps.

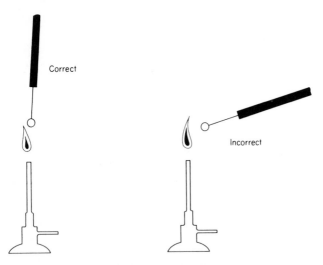

Fig. 9-4. Correct flaming of bacterial loop.

With just the pilot light going, accidental spillage of alcohol is not quite as serious as it might otherwise be. The pilot light *must* be out, however, when ether is used.

4. Filtration

Filtration is a method by which particles of various types can be separated from a medium, either liquid or gas. The filter itself is a porous material through which the medium passes, leaving behind those particles too large to pass through the pores.

There are many different filters designed for different systems, but essentially there are two basic purposes of filtration, namely, retention and separation. In retention filtration, a filter retains wanted particles from the medium and permits passage of the unwanted material through the filter. The separation filtration method allows the passage of wanted material through the filter while unwanted material is retained. It is the latter type which is of most concern in bacteriology and virology.

Occasionally, it is necessary to sterilize liquids which cannot be subjected to heat: serum, sugar solutions, enzymes, etc. For these, filtration is the method of choice. Only two of the many available filter types will be discussed: asbestos filters and membrane filters.

a. SEITZ FILTERS

Seitz filters consist of cellulose that incorporates purified asbestos (the filtering component) pressed into pad form. Several pore sizes are available (Table 9-2). If you could see a filter on cross section, it would be noted that it is graduated in density and has a direction of maximal flow.

Two types of Seitz filters allow for pressure or vacuum filtration of small volumes. Units for double and triple filtration are also available.

TABLE 9-2
Description and Properties of Seitz* Filter Pads

Designation	Flow rate (gals/sq ft/hr) at 3 lb pressure	Pore size (μ)
K2	150	5
K3	100	3–4
K5	50	2
K7	40	1–2
GP	30	1
S	15	0.5
EK	10	0.1
S1	7.5	0.05
S3	5	0.01

*Republic Seitz Filter Corp., Milldale, Conn. 06467

b. MEMBRANE FILTERS

These filters are made of cellulose esters, nylon, polyvinyl chloride, etc., in thin, porous, circular membranes and are available in several different grades of porosity and diameters (Table 9-3). The amount of fluid to be filtered will determine the scale of operation.

For volumes less than 10 ml, a Swinney adaptor attached to a hypodermic syringe is used. The membrane is held within the adaptor unit. Volumes from 10 ml to 1 liter and preparations which are too viscous to be processed through a Swinney adaptor can be processed through one of several units, either a membrane held on a fritted glass disc or in a stainless steel assembly. The first type employs vacuum and the latter uses either vacuum or pressure for filtration.

TABLE 9-3

Description and Properties of Millipore* Membranes

Designation	Mean pore size (μm)	Bubbling pressure (psi)	Water flow rate (ml/min/sq cm)†	Autoclavability
SC	8.0	4	850	Yes
SM	5.0	6	540	Yes
SS	3.0	10	400	Yes
RA	1.2	12	300	Yes
AA	0.8	16	212	Yes
DA	0.65	21	150	Yes
HA	0.45	33	52	Yes
PH	0.30	40	40	Yes
GS	0.22	55	21	Yes
VC	0.10	250	2	Yes
VM	0.05	375	1	Yes
VS	0.025	500	0.2	Yes

*Millipore Corp., Bedford, Mass.
†At 25°C, 13.5 psi.

5. Ultraviolet Light

Ultraviolet (UV) light consists of wave lengths from around 4000 to 136 Å. Those most useful in biology range between 2800 and 2300 Å, especially 2537 Å. UV light has only slight powers of penetration and is effective mainly on surfaces. Preferably, the laboratory area should be provided with radiating tubes giving off light at 2537 Å. These should be hanging from the ceiling and should be supplied with paneling to bounce the light off the ceiling (to avoid burns to the face and hands of personnel working beneath).

Ultraviolet cabinets may be commercially obtained wherein you can place articles of a solid nature which cannot be heat-sterilized (filter pads, plastic ware, for instance). These cabinets have a sliding glass door so that you can work in front of the cabinet. Always wear safety glasses (if you do not normally wear glasses) and keep the sliding glass front of the cabinet as far in the down position as will allow arms to be inside the cabinet. Wear these glasses even though you are near the light for only a few seconds. Anyone who has suffered from exposure to UV light (one of the authors included) can attest to the wisdom of this rule. If the glasses do not have shields which touch the face around the eyes, place cotton around the glasses, particularly in the area across the nose, under the eyes, and around the sides of the eyes under the ear pieces of the glasses. This will block UV rays which can enter around the glass frames.

Wear a long-sleeved lab coat or protective arm shields, particularly if you plan to be working at the cabinet for longer than 1 or 2 min. Severe burns can occur in an extremely short length of time under these intense UV lights. This is one time when looks must take a back seat to common sense!

6. Sterilizing Gases

Sterilizing gases (sterilants) affect only exposed surfaces or those materials which are porous or permeable. They have a powerful germicidal effect on bacteria and viruses. Instruments, filters, plastics, rubber goods, and even some liquids can be sterilized with gas. They have the advantage that heat and aqueous solutions are not needed, and the disadvantages that controlled temperature and humidity are required; they are toxic and irritating; and some are explosive, polymerize on surfaces, or may be unstable.

Ethylene oxide (CH_2CH_2O), probably the most commonly used sterilant today, is highly toxic and irritating. It is explosive and is, therefore, mixed with carbon dioxide or other diluents. Commercially available gases include Cryoxide (11 % plus 89 % halogenated petroleum), Carboxide (10 % plus 90 % CO_2), and Oxyfume (20 % plus 80 % CO_2). Which one you use will depend on the situation.

Commercially available gas apparatus includes a standard size autoclave in which humidity, temperature, pressure, and gas are controlled automatically, and small, portable, electrically operated units which use Freon, a non-flammable, highly penetrable gas. These units generally operate around 58°C (125 to 139°F) and 40 % humidity. For sterilization, 450 to 1000 mg of gas per liter are necessary.

G.   DISINFECTANTS

There are many disinfectants on the market. Choose a non-irritating, almost odorless, non-staining one. The presence of a surfactant permits slow release of the active portion and lowers the surface tension of the solution.

The amount of disinfectant to be used in a working solution will be stated on the container. Follow manufacturer's directions carefully and change the solutions frequently. Where commercial products are not readily available, the standard disinfectants used in bacteriology may be employed.

Iodophors are non-staining, non-irritating, almost odorless organic compounds of iodine combined with a surfactant. Iodophors lower the tension of the solution and release iodine slowly. Wescodyne, Betadine, and Ioclide are examples of commercially available iodophors used for disinfecting purposes. In solution, these are a deep, reddish brown, and the amount of iodine left in solution can be judged by comparing the amount of color remaining with that of the full strength solution. The common aqueous solution contains approximately 200 parts per million iodine; depending on the concentration and temperature, iodophors are effective against molds, bacteria, spores, and some viruses.

There are many types of phenolic disinfectants which are surfactants in themselves. In combination with detergents important as disinfectants are Roccal, Zephiran, and Phemerol Chloride, quarternary ammonium salts in which the surfactant is a cation. Its bipolar nature makes a "quarternary" an emulsifying agent for lipids and, as such, appears to be effective against viruses with lipid-rich components. Its action against other viruses is questionable, however. Quarternary detergents are not sporicidal but tinctures (alcoholic solutions) are effective against some molds and are very effective against tubercle bacilli. Cationic quarternaries should never be mixed with anionics (soaps, for example) since the ions have incompatible charges and the effects of each are thereby canceled.

The type of disinfectant you use must be dictated not only by the surface or tissue to be treated but, more importantly, by the organisms or viruses you are using. You should make a habit of washing down work areas with the appropriate disinfectant before and after use. Wash down walls, the surfaces of test tube racks, the outside of flasks, or any other hard surface area.

H.   PREPARATION OF MATERIALS FOR STERILIZATION

The type of materials to be sterilized will determine the method (moist heat, dry heat, ultraviolet light, filtration) to be employed (Table 9-4). Where equipment and glassware are used, wrapping and/or packing into containers before sterilization is necessary. Such materials are then ready for shelf storage immediately thereafter. Should budgets or other factors not permit the use of aluminum foil, heavy wrapping paper may be used provided it is not waxed or impregnated with water-soluble dyes. Foil needs only to be pressed into place to obtain a good seal. Paper may be tied with a string or taped. *Do not permit*

TABLE 9-4

Guide to Methods of Sterilization*

Article	Dry heat	Moist heat	Method: Flame	Ultraviolet	Filtration
Aluminum foil	+1	+1	+1		
Brushes		+1			
Cornwall syringes		+1			
Filter pads and discs	Note manufacturer's directions. Some may be moist-heat sterilized.				
		+1		+1	
Glassware: Flasks, pipets, test tubes	+1	+2			
Instruments (metal): Scissors, forceps	+2	+1	+3		
Media, liquids, distilled water		+1			+1
Media, liquids (with protein or sugar)					+1
Needles	+2	+1			
Plasticware	Note manufacturer's directions. Some may be moist-heat sterilized.				
Plastic-covered magnetic stirring bars		+1		+2	
Plastic caps and tubes		+1		+2	
Rubber items					
Gloves		+1			
Stoppers		+1			
Tubing		+1			
Syringes	+2	+1			
Talcum powder	+1				
Tissue grinders	+1				
Vaseline (petroleum jelly)	+1				

*Numbers given indicate suggested order of preference.

tape to stick to the glass because the heat of sterilization will make it almost impossible to remove. Place all items in baskets for ease of handling. Sterilize appropriately.

1. Glassware

After appropriate wrapping and packing (given below), glassware (test tubes, graduate cylinders, funnels, pipets, etc.) can be sterilized at 160°C

(325°F) for 2 hr under dry heat (4 hr if the oven is full), or at 15 lb pressure for 15 min under moist heat. The former is the preferred method.

a.   UNSTOPPERED TEST TUBES

These may be mass packed in metal or cardboard containers (cans, boxes) of a suitable size or may be individually wrapped and sterilized. Where large numbers of tubes are to be filled with fluids (media, cell cultures), pack them mouth up in neat rows to facilitate the filling process later. Otherwise, pack tubes, mouth down, in unwaxed quart ice cream containers. This will permit the aseptic removal of one or two tubes from a pack arrangement, while maintaining the sterility of the untouched tubes. Cover the open end of the container with a double layer of foil. After sterilization, overwrap with paper to serve as a dust protector during shelf storage.

b.   STOPPERED TEST TUBES

Sterilization times and temperatures will be dictated by the contents of the stoppered tubes. Where tubes are being sterilized empty for later use, they may be packed in baskets, and the top of the basket covered with foil or wrapping paper. The tubes will then be ready for shelf storage immediately afterward. Sterilize empty stoppered tubes at 160°C (325°F) for 2 hr with dry heat (4 hr if the oven is full) or in an autoclave at 15 lb pressure for 15 min. The former is the preferred method.

(1) *Cotton Stoppers* Stoppers made of cotton (no gauze overwrap) are usually used once only and then discarded. They can be used a number of times if well made but where extended reuse is anticipated, gauze-covered cotton stoppers should be considered, (2) below. Cotton for stoppers is usually supplied in two forms: as rolled sheets or in coils. The rolled cotton sheets come in 12- to 15-in. widths in lengths of 12 ft or more. The coiled cotton can be purchased in several thicknesses and is packed as a continuous roping in large drums or cases. In either case, be sure that the cotton you use is non-absorbent and *not* absorbent cotton. The use of absorbent cotton would defeat the whole purpose of using a stopper in the first place!

(a) *Disposable type* Cut an appropriate length of cotton from a coil supply or a square from the sheet supply. The dimensions for either one should be such that, after being pushed down into the mouth of the container, there is still enough to protrude at least 1 in. above the lip. The thickness of the rolled cotton may require its being separated into two thinner square layers [Fig. 9-5(a), (b)]. Place a square of cotton on a flat surface and roll it

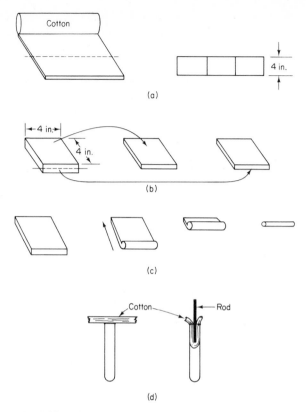

Fig. 9-5. Making disposable cotton stoppers.

up [Fig. 9-5(c)]. You should now have a length of cotton similar to a strip from the coil supply.

   (i) Place the length of coil or rolled cotton across the mouth of the tube or flask. Push the cotton down into the mouth of the container, leaving about 1 in. above the lip to grasp when removing the stopper [Fig. 9-5(d)].

   (ii) Alternatively, grasp the center of the cotton strip with forceps. Push the cotton into the tube. Before releasing the cotton, jerk upward a bit with the forceps and then release them. This will "set" the cotton.

(b) *Reusable type*   These stoppers are called *reusable* because, in making them, the twisting of the coiled cotton compacts the fibers. Autoclaving sets the form of the stoppers and makes possible their repeated use.

   (i) Grasp a piece of coiled, rope cotton between the thumb and forefinger of your right hand. With the coil in your left hand, run the length of the coil backward, over the knuckles of your right hand, then foreward under the palm of the hand, and overlap the end of the cotton held between the thumb and forefinger. Break the length of cotton off with an $1\frac{1}{2}$-in. overlap [Fig. 9-6(a)–(c)].

(ii) Without breaking your grasp on the cotton, slip the circlet of cotton off your knuckles. Flatten the circle by grasping

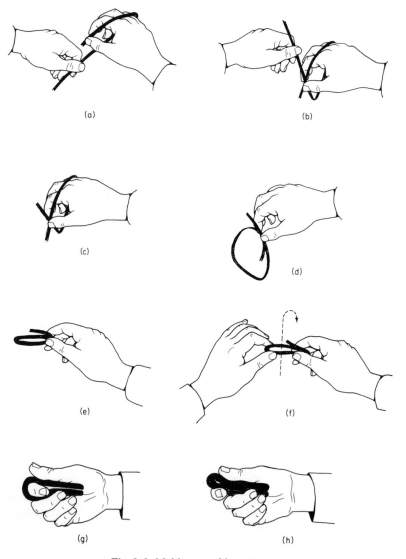

Fig. 9-6. Making reusable cotton stoppers.

the opposite side of the circle and press it against the overlap held between your thumb and forefinger [Fig. 9-6(d), (e)].

(iii) With the left hand, grasp the left-hand end of the cotton and, still holding the overlap and center end compressed by the right hand, twist the left-hand side one-quarter turn away from you [Fig. 9-6(f)].

(iv) Without releasing the left hand, continue by folding the twisted portion backward toward your wrist until the center portion is wrapped over your index finger and you now have two parallel lengths of cotton [Fig. 9-6(g)].

(v) Grasp the twisted and folded-back portion with your ring and little fingers, compressing the parallel sides together while simultaneously slipping your forefinger out from the fold [Fig. 9-6(h)].

If you have done this properly, you should have the original overlap to the inside, the outside of the plug should be smooth, and the ends at the top should be smooth. The plug should fit into a test tube with ease. After sterilization, the plug will become compacted, can be removed and replaced in the tube with ease without opening up, and should be reusable many times before it has to be discarded. The following is a good test for the adequacy of the stopper: (1) Insert it into a tube and then grasp the stopper. If properly made, you should be able to lift the tube from the bench top. (2) After removing the stopper from the tube, you should be able to insert it a number of times without its losing its shape.

(2)   *Gauze-covered Type*   The following method may be used to make any size stopper from small glass needle holder size to those required for large flasks. When properly made, such stoppers can withstand repeated use.

(a) Cut out a double thickness of gauze such that the distance from any side is twice the length of the plug desired. Place the gauze over the mouth of the container for which a stopper is being made and push the center downward with a round-bottomed glass rod. Avoid a sharp instrument since it will penetrate the material separating, if not cutting, the fibers [Fig. 9-7(a), (b)].

(b) Pack small pieces of cotton into the gauze "cup" until it is filled. A compact filling will be achieved if the gauze is pulled upward around the lip of the container with one hand while the cotton is tamped into place with the other. Continue this packing technique as more cotton is added while you rotate the flask in a counterclockwise direction [Fig. 9-7(c)]. The idea is to work around the stopper, pulling the gauze up tightly to avoid any creases or folds which might act as air passageways when the stopper is finished and in place.

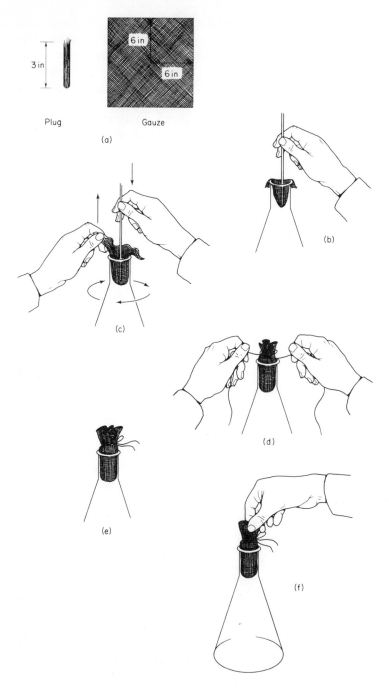

Fig. 9-7. Making gauze-covered stoppers.

(c) When the stopper is solidly packed, the gauze is smooth, and the stopper is as deep as desired, tie the gauze tightly with string, just above the cotton [Fig. 9-7(d), (e)]. Cut off the excess gauze approximately $\frac{1}{2}$ in. above the string.

(d) If packed tightly and tied off close to the stopper, the plug should remain firm. If done properly, you should be able to pick up the flask by grasping the top of the plug [Fig. 9-7(f)].

c. FLASKS, GRADUATE CYLINDERS, WIDE-MOUTHED CONTAINERS

Cover the mouth of this type of glassware with double layers of foil. Flasks may also be closed with a gauze-covered cotton plug (above) and overwrapped with foil (or a double layer of wrapping paper, tied in place with string).

d. FUNNELS

Place double folds of foil over the stem end. Put the funnel, open end down, in the center of a sheet of wrap (preferably foil) which is sufficiently large to permit its being brought up over the stem and secured. Wrapped in this manner, the funnel can be set in place after removing the stem wrap, while the remaining wrap serves to cover the mouth and maintain sterility.

e. PIPETS

Plug all pipets with longer-fiber, non-absorbent cotton. Be sure you do *not* use absorbent cotton! If you do not have an automatic plugging machine, the pipets may be plugged manually as follows:

(1) Open up a roll of non-absorbent cotton.

(2) Open out one end of an ordinary paper clip [Fig. 9-8(a)]—the other end is your handle. Brush the single end of the clip lightly over the surface of the cotton, picking up only a very small amount [Fig. 9-8(b)]. Transfer the cotton to the mouth of the pipet, using the clip to lightly pack the cotton into the mouthpiece. Do not push the cotton beyond the constriction [Fig. 9-8(c)]. Any wisps of cotton remaining (which are extremely annoying when you are pipetting) should be eliminated by taking a bundle of plugged pipets in your hands and quickly passing the mouths through a Bunsen burner flame.

(3) Trial and error will tell you how much cotton to pack into the pipet. There should be enough to inhibit the passage of materials through and around the plug but it should not be packed so tightly that suction is difficult to attain.

(a)

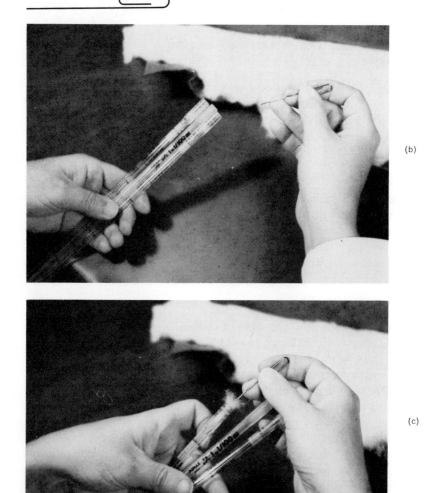

(b)

(c)

Fig. 9-8. Stuffing pipets.

(4) Packing of Pipets for Sterilization—Group Pack:

   (a) Place a wad of cotton or several layers of gauze on the bottom of a pipet canister to act as a bumper pad. Put the pipets into the canister, pointed ends first, so the points rest on the bumper pad. After sterilization, you will be able to remove a pipet by grasping the exposed end, while the tip remains untouched.

   (b) Mark the cap end of the canister with an indelible ink to indicate the pipet size. Wax pencil or paper labels will not resist the heat. Color-coded canisters are helpful.

   (c) Place the cover on the canister but do not close it. Leave a $\frac{1}{4}$-in. opening between the top and bottom to facilitate heat circulation within the canister. (Pipets may be autoclaved but because of the moisture which condenses on them, long drying times are necessary afterward, making this a less desirable method.) Place the canisters in an oven and sterilize at 160°C (325°F) for 3 to 4 hr. At the end of the sterilization period, open the oven door and, using asbestos gloves, quickly close the pipet canisters.

(5) Packing of Pipets for Sterilization—Individual Wrap: If you do not have pipet canisters or if you have large pipets (25 ml, for example), very large glass test tubes or long mailing tubes can be substituted. Alternatively, pipets may be individually wrapped as follows:

   (a) Cut a square of paper such that the diagonal measurement is 4 in. longer than the length of the pipet [Fig. 9-9(a)]. Cut the paper diagonally in half [Fig. 9-9(b)]. Place the lower half of the pipet on the edge of the left-hand piece of paper [Fig. 9-9(c)] and roll the paper and pipet away from you until the pipet is in line with the left-hand corner of the paper [Fig. 9-9(d)]. Fold in the corner [Fig. 9-9(e)] and continue rolling until the far edge is reached. Seal the paper with tape.

   (b) Wrap the top half of the pipet in the right-hand piece of paper with the diagonal cut edge of the paper overlapping the wrapped end of the pipet by 1 in. or more [Fig. 9-9-(f)]. Do not seal the top wrapper *to* the bottom wrapper. (After sterilization you want to be able to slip off the top covering in order to grasp your pipet, while the bottom of the pipet remains protected. The tip end can be slipped back into the wrapper to keep the pipet sterile if it is to be reused.) Sterilize.

2. Rubber Stoppers and Plastic Caps

   a. Place rubber stoppers or plastic caps in unwaxed ice cream containers. Large-mouthed glass jars with screw caps, coffee cans,

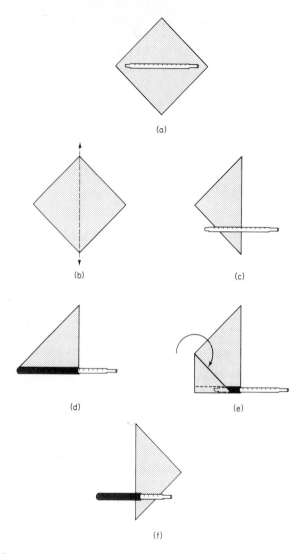

Fig. 9-9. Wrapping a pipet.

etc. will also serve the purpose. *Do not pack the stoppers tightly. Leave head room.* Mark the lids with pencil, indicating color, size, and type of stopper.

b. Place in wire baskets and autoclave at 15 lb pressure, 15 min. Shake the containers vigorously at intervals during the cooling period to prevent the stoppers from sticking together. The stoppers may be stored in these containers where they will remain sterile, and the containers may be reused many times.

3. Syringes and Needles

a. Assemble syringes, being sure to match the plungers to the barrels (i.e., both will have the same number printed on them) unless you are using universal-fit syringes.

b. Place a bumper pad of cotton or gauze in the bottom of test tubes large enough to accommodate the particular syringe or needle you may have.

c. Place the syringe into the tube, needle-holding end (Luer-Lok) down and plunger end up. Put needles in, point down. Close with a gauze-covered cotton plug (Fig. 9-7). Alternatively, syringes and needles may be wrapped in aluminum foil or wrapping paper. Sterilize at 160°C (325°F) for 2 hr.

4. Cornwall Syringes

a. Assemble the various parts in the order shown in Fig. 9-10(a), (b). Be sure the various small parts of the automatic valve are present and in place. Replacement parts are available from the manufacturer if you should lose the springs or tack valves.

b. Lock the smaller end of the two-way valve to the bottom of the syringe. This requires only a single twist to lock the two together.

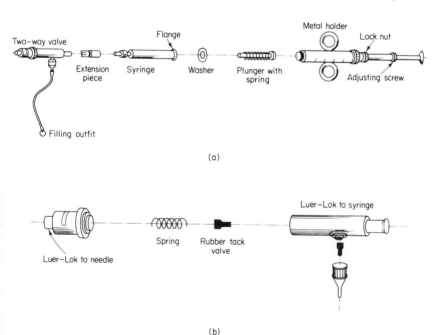

Fig. 9-10. Assembling a Cornwall syringe.

    c. Fold a square of aluminum foil over the metal sinker ball. The whole unit should now be complete except for a needle which you will add when ready to use the Cornwall.

    d. Wrap the Cornwall syringe as shown in Fig. 9-11(a)–(i). The finished product should look like a mailing envelope with the flap tucked inside. It is better to use this fold-in technique to close the package than to seal it with tape. After sterilization, tape is most difficult to remove, and often it becomes necessary

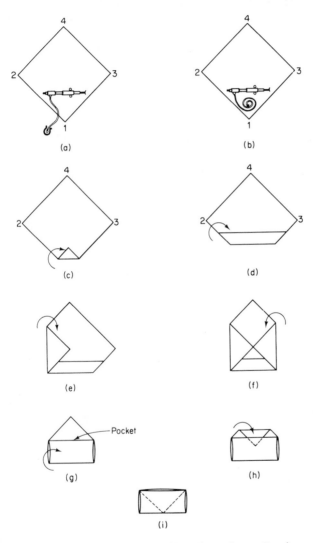

Fig. 9-11. Wrapping a Cornwall syringe.

to tear the package open, exposing the Cornwall to contamination. It is simpler and easier merely to unfold the package and still maintain sterility.

e. Place in wire baskets and autoclave at 15 lb for 20 min.

5. Gauze Squares

Sterile gauze squares may be obtained commercially in various sizes, or you can wrap unsterile ones in packets of one or more in aluminum foil and autoclave at 15 lb pressure for 20 min. It may be necessary to make your own gauze squares. All you need is a diagram drawn on cardboard, scissors, 12 ply (U.S.P. Type VII) gauze, and a metal ruler. Proceed as follows:

a. On a piece of cardboard, draw a diagram which is three times longer and wider than the size of your finished square. Add at least an additional $\frac{1}{4}$ in. all around [Fig. 9-12(a)].

EXAMPLE

You want a 3×3 in. finished square. Make a $9\frac{1}{4} \times 9\frac{1}{4}$ in. diagram on cardboard.

b. Place a piece of gauze cut to this size on top of the cardboard diagram. Fold each $\frac{1}{4}$-in. edge of gauze inward and down onto

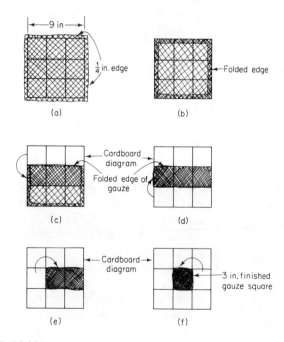

Fig. 9-12. Making gauze squares.

the large area of the gauze and run the ruler along the edges to press the $\frac{1}{4}$-in. folds into place [Fig. 9-12(b)]. You now have an exact 9-in. square piece. Note that the cardboard diagram is *not* incorporated into the folds of the gauze. It is merely a guide.

c. Pressing each of the layers down securely as they are made, proceed with the folding as follows: Fold the top 3 × 9 in. strip down over the center 3 × 9 in. section [Fig. 9-12(c)]; then fold the bottom 3 × 9 in. strip up over the center 3 × 9 in. section [Fig. 9-12(d)]. You now have a 3 × 9 in. piece of gauze, three layers thick.

d. Fold the left- and right-hand 3-in. pieces in over the center and press down firmly with the ruler [Figs. 9-12(e), (f)]. You now have a 3-in. finished square occupying the exact center of the cardboard diagram. Wrap and sterilize as cited above.

6. Magnetic Stirring Bars

a. Wrap in aluminum foil or place in small glass screw-capped vials. Do not tighten screw caps until you remove the containers from the autoclave because collapse of the containers may occur as they cool.

b. Autoclave at 15 lb pressure for 15 min.

c. Tighten caps as containers cool.

7. Tissue Grinders

The two parts of a tissue grinder may be wrapped separately in aluminum foil if they are of universal fit. Most tissue grinders are matched pieces, however, and it is preferable to keep them together. They may be processed by placing them on a large square of aluminum foil or wrapping paper as follows:

a. Place the top part (pestle) of the tissue grinder in one corner of a paper square with the handle to the right [Fig. 9-13(a)].

b. Fold point No. 1 up over this piece of glassware [Fig. 9-13(b)].

c. Cover the mouth of the bottom part (mortar) of the tissue grinder with a tight-fitting piece of aluminum foil.

d. Place the mortar on the paper with the bottom to the right and the covered end to the left and up tight to the overwrapped pestle piece [Fig. 9-13(c)].

e. Fold end No. 1 up over the mortar. You will now have both pieces of the grinder overwrapped in one corner of the square [Fig. 9-13(d)].

Note that the piece of the grinder to be used first is the last to be wrapped. In this way, the mortar may be removed for filling (with its sterile

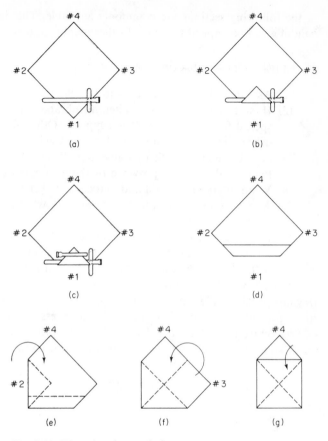

Fig. 9-13. Wrapping tissue grinders.

aluminum foil cover) while the pestle remains covered and sterile until it is required.

    f. Turn point No. 2 on the left side over to the right [Fig. 9-13(e)].

    g. Turn point No. 3 on the right over to the left [Fig. 9-13(f)].

    h. Fold point No. 1 toward point No. 4 once more. Fold point No. 4 into the envelope pocket [Fig. 9-13(g)].

8. Filter Units

Filter units of various types are supplied by several manufacturers and descriptive literature which usually accompanies the units should be carefully followed. Some membrane filters may be autoclaved, whereas others must be obtained from the manufacturer in presterilized packages. It is extremely important that you realize that some membrane filters cannot tolerate any sterilization procedures which utilize heat, wet or dry. The units described in the following sections are commonly available. The description of their handling can be applied to units of other manufacturers.

a.   FILTER UNITS—MEMBRANE—SWINNEY UNIT

(1) Assemble the unit according to Fig. 9-14.

(2) Handle the autoclavable filter membrane very carefully using the special flat-mouthed forceps supplied. Other forceps will tear the delicate membrane and should not be used.

(3) Secure the entire unit by tightening all threaded components using the special wrenches provided by the manufacturer.

(4) Wrap in aluminum foil and autoclave at 121°C (250°F) for 15 min.

(5) Allow the autoclave pressure to drop slowly to preserve the membrane structure.

(6) If the particular membrane is not autoclavable (Table 9-3), process the unit without it and add a commercially sterilized membrane, with sterile technique, at the time the unit is used.

Filter units, both non-sterile and sterile, are also available in plastic (Swinnex). These are reusable, so follow manufacturer's directions. New units, presealed with the filter and the holder pressed into a single unit, can be used only once and must be discarded. The units are excellent for sterilizing small volumes of dye solutions, serum, and tissue culture media.

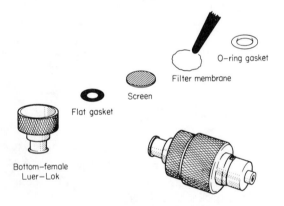

Fig. 9-14. Swinney filter unit.

b.   FILTER UNITS—MEMBRANE—ALL GLASS UNIT

(1) Assemble the unit according to Fig. 9-15.

(2) Wrap in aluminum foil.

(3) Wrap the side arm, and the top of a sidearm Erlenmeyer flask with aluminum foil.

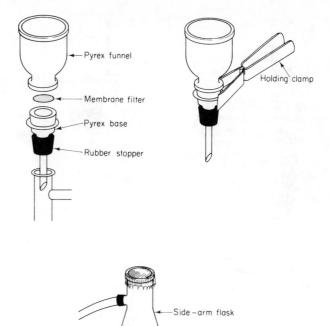

Fig. 9-15. Filter unit, membrane, glass.

(4) Autoclave the flask and filter units at 121°C (250°F) for a minimum of 30 min and a maximum of 45 min.

c. FILTER UNITS—MEMBRANE [FIG. 9-16(a)] OR SEITZ
TYPE [FIG. 9-16(b)]

    (1) Assemble the unit according to Fig. 9-16(a) or (b).
    (2) Wrap in aluminum foil and autoclave as above.

d. FILTER UNITS—SEITZ—PRESSURE UNIT, LARGE
VOLUME (FIG. 9-17)

    (1) Assemble the unit.
    (2) Remove the pressure gauge and pack the filter outlet with cotton.
    (3) Wrap the glass bell in aluminum foil.

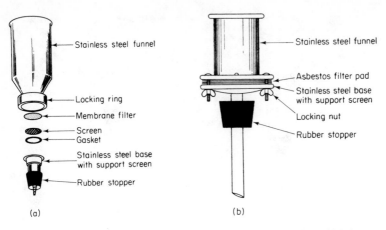

Fig. 9-16. (a) Filter unit, membrane, Seitz; (b) Seitz vacuum unit.

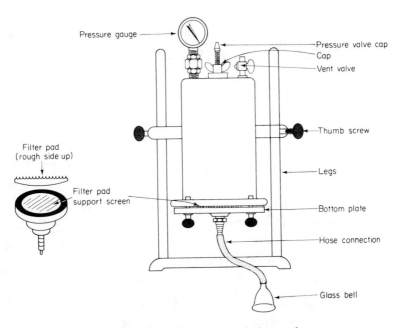

Fig. 9-17. Seitz pressure unit, large volume.

(4) Carefully remove the heavy unit from the support stand, and place it in a wire basket. Autoclave at 18 lb pressure (125°C) for 30 min.

e.   FILTER UNITS—SEITZ—PRESSURE UNIT, SMALL
VOLUME

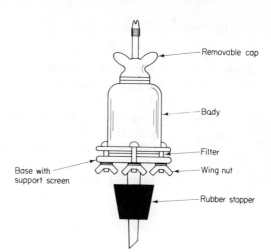

Removable cap

Body

Filter

Base with
support screen

Wing nut

Rubber stopper

Fig. 9-18. Seitz pressure unit, small volume.

(1) Assemble the unit according to Fig. 9-18.

(2) Wrap in aluminum foil and autoclave at 125°C (18 lb) for 30 min.

9. Filter Pads

Extra supplies of membrane and Seitz filter pads may be autoclaved at 15 lb pressure for 15 min if they are placed in Petri dishes. Certain filters are autoclavable (Tables 9-2 and 9-3). Other types may be purchased in sterile packets or may be sterilized by exposing them to ethylene oxide (see manufacturer's directions) or ultraviolet light.

a. Place individual filters on a sheet of foil inside the UV cabinet.

b. Leave a sterile Petri dish and a pair of flat-tipped forceps there also.

c. Pull the window into place and *then* turn on the UV light.

d. Leave the filters exposed to the UV light for several hours. The exact time will depend on the distance between the lamp and the filters and the wavelength of the UV lamp.

e. Put on a pair of safety glasses; *then* raise the glass door only far enough to give access to the cabinet.

f. Turn the filters over using the forceps provided.

g. Close the window and leave the filters exposed to the UV light for an additional number of hours.

h. Put on a pair of safety glasses; raise the window only far enough to give access to the cabinet.

i. Pick up the filters with the forceps provided and place them in the sterile Petri dish.

j. Turn off the UV light.

10. Liquids

Distilled water, culture media, salt solutions, and other liquids not destroyed by heat may be sterilized in an autoclave.

 a. Fill the tube or flask no more than two-thirds full to prevent boilover under pressure.

 b. If many tubes are to be filled, a convenient setup with ring stand and funnel (Fig. 9-24, p. 228) may be used.

 c. Plug tubes with cotton; cover flasks tightly with foil or use a gauze-covered plug and overwrap with foil or wrapping paper to prevent wetting in the autoclave. Otherwise you must be sure to remove the materials from the autoclave immediately so that the heat will dry the plugs.

 d. Autoclave at 15 lb pressure (121°C), 15 to 20 min; small volumes (10 ml) at 10 lb pressure (115°C), 12 min; large volumes (1 liter or more) at 15 to 18 lb, 20 to 30 min.

I. FILTRATION TECHNIQUES

1. Swinney

This filter system is used primarily for small volumes (2 to 30 ml) of materials, either to sterilize them or, in the case of viral suspensions, as a means of estimating viral size (i.e., by the sequential filtration of infected fluids through filters of smaller and smaller porosities).

The Swinney unit, of stainless steel, will attach directly to plastic syringes or to any glass syringe fitted with a Luer-Lok-type outlet.

a. MATERIALS

    Material to be filtered
    Syringe of appropriate size
    Needle, of proper size, sterile
    Swinney unit, assembled with proper porosity filter, sterile
    Collecting vessel, sterile

b. METHOD

 (1) With sterile technique, attach a needle to the male Luer-Lok (long end) of the Swinney, but leave the needle in its holder to keep it sterile. Do not attach the Swinney to the syringe. Set aside.

 (2) Draw the fluid to be filtered into the syringe by immersing the tip end of the syringe in the fluid and draw out the plunger (Fig. 9-19).

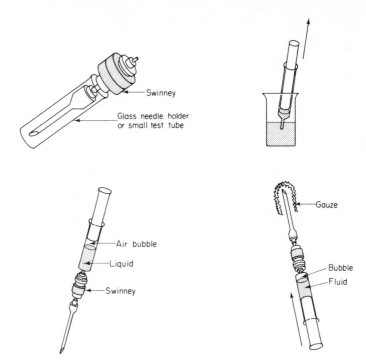

Fig. 9-19. Swinney filtration of non-viral fluids.

Note: If the material to be filtered contains virus, remove the plunger from the syringe, attach the syringe to the female Luer-Lok of the Swinney, pour the material to be filtered into the barrel of the syringe, and then replace the plunger [Fig. 9-20].

(3) Now aseptically attach the syringe to the female Luer-Lok of the Swinney.

(4) Invert the reassembled syringe (needle pointing upward).

(5) Place a sterile gauze pad over the needle.

(6) Slowly push the plunger into the barrel until the air bubble, inevitably trapped inside, is squeezed out through the filter pad. Do *not* draw the plunger *back* with the Swinney unit attached because back pressure will rupture the filter pad.

(7) With sterile technique, insert the needle well into the receiving flask. A flamed piece of aluminum foil, covering the mouth of the flask and the needle, will preclude contamination by airborne organisms. Alternatively, insert the needle through several layers of sterile gauze held over the lip of the flask.

(8) Gently compress the plunger. Expect a high degree of resistance.

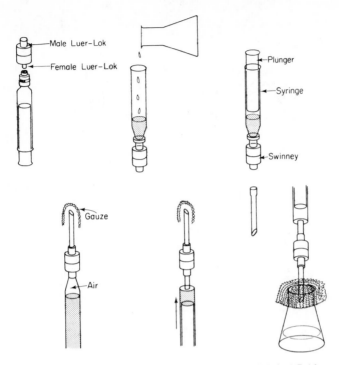

Fig. 9-20. Swinney filtration of viral fluids.

The filter area itself is limited and, where small porosities are concerned, it will take a great amount of time for the liquid to pass through the filter.

(9) With sterile technique, cover the receiving flask, label, and refrigerate or freeze the material.

(10) Remove the needle and Swinney from the syringe, and place them in a contamination pan.

(11) Hold the Luer-Lok end of the syringe in disinfectant and draw some up into the syringe. Lower the syringe into the contamination pan.

(12) Sterilize, reclean according to manufacturer's directions, add a new filter, and resterilize.

2. Glass Filter, Vacuum Type

Vacuum sterilization is normally used where large volumes of liquids must be processed and where the nature of the filter itself permits suction upon its surface. There are several types of filters which may be used, including the older clay filters, the Seitz, and the membrane. Whichever one is used, the principle is identical. Since foaming may occur during this type of filtration, it should *not* be used with fluids whose protein content is of significance to the investigation.

The glass filter, depending on size, may be used for volumes ranging from 5 to 50 ml (micro-analysis filter) to a liter (all-glass filter).

a. MATERIALS

> Material to be filtered
> Filter containing proper filter pad and rubber stopper over stem end, sterile
> Sidearm Erlenmeyer, with mouth opening to accommodate rubber stopper on filter unit above, and with non-absorbent cotton plug in side arm, sterile
> Pressure tubing, sterile
> Erlenmeyer flask, with two-hole rubber stopper with one short bent-glass tube and one long bent-glass tube (overflow flask)
> Pump, or water suction unit on faucet
> Pipets, sterile
> Test tubes, sterile

b. METHOD

(1) *Filtrate*
   (a) Screw suction unit to faucet (if pump is not used). Attach one length of pressure tubing between the side arm of the faucet unit (or the pump) and the long, bent-glass tube of the overflow flask. Attach a second length of pressure tubing between the short, bent-glass tubing of the overflow flask and the cotton-stoppered side arm of the filtration flask (Fig. 9-21).

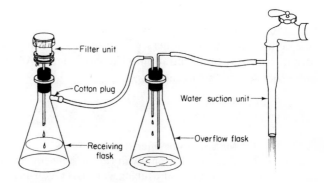

Fig. 9-21. Water suction filtration unit.

(b) With sterile technique, remove the covers from the stopper-end of the FILTER unit, and the filtration flask.

(c) Quickly place the filter unit into the mouth of the filtration flask. Leave the cover on the filter itself until ready to add the solution to be filtered.

(d) With sterile technique, add solution to the filter cup and replace the foil cover.

(e) *Slowly* turn on the water (or slowly increase the pressure if a pump is used).

(f) Add more solution to the filter cup as the liquid level goes down. Always add more sample while the funnel is still partly full in order to avoid disturbing the distribution of contaminant on the filter. When the liquid level drops to the area of the filter pad, turn off the suction.

(g) With sterile technique, remove the filter unit, and place a sterile cover over the filtration flask.

(h) With sterile technique, pour or pipet the filtrate into storage tubes. Label, and refrigerate or freeze filtrate as required.

(2) *Residue*   If, for any reason, you do *not* wish the filtrate, but want the particulate residue instead, follow the procedure above through step f.

(g) Continue running the vacuum pressure until all liquid has disappeared. Turn off suction.

(h) With sterile forceps, carefully lift the filter pad and place it into a sterile Petri dish. Label. If perfectly dry residue is desired, place the Petri dish into a desiccator and allow the material to dry over calcium chloride. Alternatively, scrape the material into a sterile round-bottom flask and process through a freeze-dry apparatus.

3. Stainless Steel Filter, Pressure Type

Pressure sterilization is used for liquids such as biologic solutions, tissue culture media, and others of a proteinaceous nature, which would be denatured by vacuum filtration. Ordinarily, large volumes (100 to 400 ml) are processed by this method.

a.   MATERIALS

Material to be filtered
Pressure filter unit with rubber stopper over exit stem end and short pressure tubing with glass tube with cotton plug on inlet stem, sterile

Sidearm Erlenmeyer flask, sterile
Sidearm Erlenmeyer flask with double-hole stopper with
long and short bent-glass tubing
Pressure tubing
Vacuum pump, pressure tank, or tire pump

b. METHOD

(1) Set up the filtration unit (Fig. 9-22). Be sure the receiving flask has
a side arm to pass air displaced by the liquid.

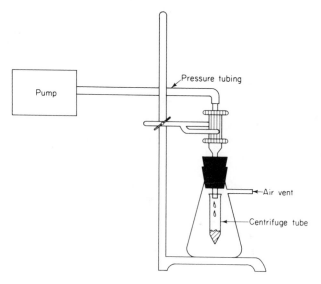

Fig. 9-22. Pressure filtration, small volume.

(2) If small volumes are to be filtered, the liquid may be placed directly
into the barrel of the filter with sterile technique; connect the pump,
via pressure tubing, to the inlet adaptor of the filter; suspend a 15-ml
centrifuge tube below the outlet by a stainless steel clamp to serve
as a collecting vessel. If large volumes are to be filtered, the set-up
shown in Fig. 9-23 is preferable.

(3) Apply pressure (maximum: 100 psi) to start filtration. Pressure may
be applied with a standard pump with tubing attached to the pressure nozzle. Alternatively, a compressed air tank may be used.
Where neither is available, a tire pump will give admirable service.
Just remove the standard hose with its metal tip and replace it with
regular tubing.

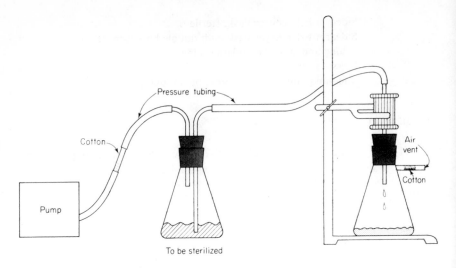

Fig. 9-23. Pressure filtration, large volume.

(4) After use, remove the filter unit and cover the receiving flask with aluminum foil (use sterile technique).

(5) Dispense the filtrate as preferred; label and date; refrigerate or freeze.

(6) Disassemble the unit; remove and discard the filter pad; clean all units in hot water and detergent; never use steel wool or abrasives on any part of the filter unit; rinse in hot running water, then cold water, and then distilled water. Allow units to air-dry. Do not dry with cloth or paper toweling since lint or fibers may contaminate the filtrates.

4. Seitz Stainless Steel Filter Unit, Pressure Type

a. MATERIALS

Materials to be filtered
Seitz stainless steel filter unit, with filter pad in place, sterile
Receiving flasks, sterile
Pump, or tire pump

b. METHOD

(1) Place the unit on its support stand at a height which will allow room for a receiving flask to go under the bell jar. Tighten the thumb

screws. Make certain all wing nuts around the base are tight. Upon application of pressure, leakage may be observed in this area. Wing nuts may then have to be tightened further with pliers.

(2) Open the refill cap and carefully pour in the liquid to be filter-sterilized. Close the refill cap.

(3) Remove the cotton from the pressure gauge opening and attach the gauge.

(4) Attach pressure hose from the pump (or attach a tire pump if an electric one is not available) to the vent valve. Be sure the vent is open.

(5) Place a sterile flask into position under the glass bell.
Note: With this method, initial filtrates will be slightly colored due to release of asbestos particles and Mg^{++} ions from the pad itself. It is customary to discard the first filtrate and start collecting only when the eluate is clear. Alternately, you can process 500 ml of sterile distilled water through the unit first.

(6) Turn on the pump and increase the pressure to 8 lb.

(7) When the flask is full, quickly replace it with another sterile flask.
Note: This will require some skill since the liquid, under pressure, is continually flowing from the filter. When a flask is almost full, remove the aluminum foil cover from an empty flask, flame the lip of the flask, remove the full flask, and hold its lip to the flame while placing the empty flask under the bell jar. Flame the foil cover and place tightly on the full flask.

(8) When all the fluid has been processed, lower the pressure and disconnect the pump.

(9) Place sterile rubber stoppers into all flasks and label, date, initial, and refrigerate the material.

(10) Dismantle the filter unit; wash the inside of the tank and the metal filter holder with detergent solution and a brush. Rinse. Dry thoroughly. Reassemble with a new filter pad and resterilize according to Section F.

J. MEDIA, GENERAL STORES

Many materials must be prepared and kept on hand either for direct use or as additives to media and other solutions. The following materials constitute a basic list and should be kept on hand at all times:

> Cardboard spatulas
> Weighing paper
> Aluminum foil
> Labeling tape

Long glass rods
Magnetic stirring bars, sterile
Balance
Bunsen burner
Pump
Pipets, various sizes, sterile
Graduate cylinders, various sizes, sterile
Erlenmeyer flasks, various sizes, sterile
Prescription bottles, various sizes, sterile
Rubber stoppers to fit above, black and white, sterile
Funnels, sterile
Seitz and membrane filters, sterile

Some of the materials will be sterilized after preparation; i.e., sterile techniques are not necessary. In other instances, you will be combining solutions which already have been sterilized. In such instances, sterile techniques must be observed. An example of proper sterile techniques is given under the section concerned with the preparation of phosphate buffered saline, and the techniques should be used when preparing any solutions involving sterile ingredients.

Clear the work area of all unnecessary materials. Wash down the work area with disinfectant to free it of any dust, accumulated chemicals, or media from previous work and to cut down the number of undesirable organisms.

All chemicals used for media preparation should be kept on a separate shelf and confined to this use only. This will ensure complete control over the quality of the material in the bottles over the shelf life of the chemical.

It is good practice to place the bottles of chemicals to be used on your left in order of use, with the first chemical closest to you and the last to be used at the farthermost position. As you use each chemical, the jar is then placed on your right, the first chemical in the farthermost position, the last chemical used in the closest position. This order may be reversed if you are left-handed. This practice helps to eliminate any question of duplication should you be interrupted during the weighing procedure. If the bottle is on the right side, you *know* you have weighed out that particular chemical (Fig. 9-24).

Make it a habit to remove from the jar only that amount of material you estimate you need. Trial and error will soon enable you to do this with

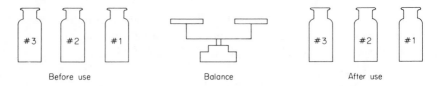

Before use   Balance   After use

Fig. 9-24. Weighing procedure.

amazing accuracy. Pour the powder into a small dish made by turning up the edges of a square of aluminum foil or other stiff paper. Keep the lid on the jar to prevent contamination and the uptake of moisture.

Get *out* of the habit of using spatulas. Far too often these are never properly washed or are only wiped with paper toweling. There is too much of a chance of cross-contaminating your materials. Instead, use a.cardboard strip or soda-straw scoop (see below) to lift the powder to the balance. For cardboard strips, cut a 3 × 5 in. file card into six $\frac{1}{2}$ in. wide strips. Fold each strip lengthwise in half, thus making an individual scoop with $\frac{1}{4}$ in. sides [Fig. 9-25(a)]. For soda-straw scoops, hold the straw firmly, cut it in half diagonally at about a 20° angle with a sharp razor blade or Bard-Parker blade [Fig. 9-25(b)].

Never put any spatula into a container of chemicals. Pour more powder out into a "dish" if you need it. When done with *each* powdered medium or chemical, discard the scoop and any powder left in the dish. If you have developed a good "measuring eye," you will discard an extremely small amount.

If you use chemicals which are hydrated or unhydrated instead of those given in any formula, see Appendix A for the method of adjusting the amount required. A hydrated chemical is one which contains water as part of its composition. These are designated in the formulas as "·X molecules H_2O", i.e., $CaCl_2 \cdot H_2O$, or $MgSO_4 \cdot 7H_2O$. Be cautious when choosing a bottle of a chemical from the reagent supply. Do not use one as a substitute for the other without making the necessary adjustments in weight.

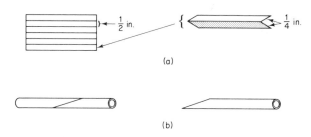

(a)

(b)

Fig. 9-25. Scoops.

1. Distilled Water

Pass triple-distilled water through a deionizer to remove excess salts which are not removed during distillation, particularly when a metal still is used. Pour the water into Erlenmeyer flasks, filling them no more than two-thirds full. Place a tight-fitting double fold of aluminum foil over the top. Autoclave at 15 lb pressure, 20 min.

2. Nutrient Agar

Nutrient agar powder	20 g
Distilled water to make	1000 ml

Suspend the agar in 950 ml of water and heat in a double boiler to dissolve. Do not use a beaker as the agar tends to stick to the bottom and the beaker may break during the heating process. Be sure that the agar does not boil vigorously because it will rapidly overflow the pan. Agar is extremely slippery and can be dangerous if it flows onto the floor. Do not attempt to wipe up hot agar with paper toweling. Instead, allow the agar to cool, at which time it will easily peel off any surface.

Pour the hot solution into a graduate cylinder and add water to the liter mark. Set up a ring stand with glass funnel and be sure the clamp is tightly in place (Fig. 9-26). Pour the agar into the funnel. Dispense in 10-ml amounts into test tubes. Plug with non-absorbent cotton plugs or add metal closures.

Autoclave at 15 lb pressure for 15 min. Slant the tubes and allow the agar to solidify. There are special metal trays for this purpose. A long strip of board, $\frac{1}{2}$ in. thick, placed on the bench top, will serve as well. Rest the plugged end of the tube on the wood strip. Store the solidified slants in a wire basket, label, and refrigerate.

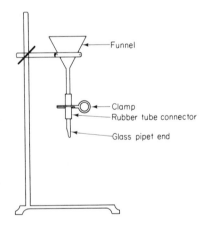

Fig. 9-26. Ring stand and funnel.

3. Top-Layer Agar

Bacto agar	7.5 g
Tryptone	2.5
Nutrient broth powder	8.0
Yeast extract	2.5
NaCl	5.0
Distilled water to make	1000.0 ml

Mix ingredients with water and dissolve with gentle heat in a pan. Pour into a graduate cylinder and add water to the liter mark. Pour into a funnel held in a ring stand. Dispense 2.5 ml into tubes, plug, and autoclave at 15 lb pressure, 15 min. Cool in upright position. Store in wire baskets, label, and refrigerate.

4. Bottom-Layer-Agar

Prepare as indicated for top-layer agar above except increase the amount of agar to 15 g.

5. Plaque Suspension Agar

Bacto agar	0.9 g
Lactalbumin hydrolysate	0.5
Yeast extract	0.1
Bovine albumin	0.1
Earle's saline to make	100.0 ml

Dissolve the ingredients in saline using gentle heat. Make up to volume with saline. Dispense in 2-ml amounts in test tubes. Autoclave at 15 lb pressure, 15 min. Cool in upright position. Store in wire baskets, label, and refrigerate.

6. Deca-Strength Phage Broth

Peptone	100.0 g
Yeast extract	50.0
NaCl	25.0
K_2HPO_4	80.0
Distilled water to make	1000.0 ml

Dissolve ingredients in enough water to make 1 liter. Adjust to pH 7.6 if required. Place in a 2-liter flask, and cover with aluminum foil. Autoclave at 15 lb pressure, 15 min. Cool and store in wire baskets, label, and refrigerate.

7. Thioglycollate Broth

Thioglycollate broth powder	29.5 g
Distilled water to make	1000.0 ml

Dissolve thioglycollate powder in 950 ml of distilled water by bringing to a boil. Make up to 1 liter by adding water. Pour into a glass funnel held in a ring stand. Dispense 15-ml amounts into test tubes. Plug with cotton stoppers or cover with metal closures. Autoclave at 15 lb pressure, 15 min. Cool and store in wire baskets, label, and refrigerate.

8. Tryptose Broth, 5.0%

Tryptose powder	50.0 g
Distilled water to make	1000.0 ml

Dissolve tryptose powder in 950 ml of water with gentle heat. Make up volume to 1 liter with water. Pour into a 2-liter flask. Add 1.0 ml of 4.0% NaOH for each 100 ml of tryptose broth to achieve a pH of 7.4. Cover with aluminum foil. Autoclave at 15 lb pressure, 15 min. Cool, label and refrigerate.

9. Tryptose Phosphate Broth

Tryptose powder	20.0 g
Dextrose	2.0
NaCl	5.0
Na_2HPO_4	2.5
Distilled water to make	1000.0 ml

Dissolve ingredients in 950 ml of distilled water with gentle heat. Make volume to 1 liter with water. Pour into a 2-liter flask and cover with foil. Autoclave at 15 lb pressure, 15 min. Cool, label, and refrigerate.

10. Lactalbumin Hydrolysate (LAH), 5.0% [35]

Lactalbumin hydrolysate powder	50.0 g
Distilled water to make	1000.0 ml

Dissolve powder in 950 ml of water with gentle heat. Make volume to 1 liter with water. Place in a 2-liter flask. Add 1.0 ml of 4.0% NaOH for each 100 ml of LAH to achieve a pH of 7.4. Cover with aluminum foil. Autoclave at 10 lb pressure, 10 min. Cool, label, and refrigerate.

11. Sodium Hydroxide, 4.0%

Sodium hydroxide pellets	4.0 g
Distilled water to make	100.0 ml

Dissolve NaOH in enough water to make 100 ml. Pour into a flask and cover with aluminum foil. Autoclave at 15 lb pressure, 15 min. Cool and plug with a sterile rubber stopper and overwrap with a foil cap. Store at room temperature or in the refrigerator.

12. Sodium Bicarbonate, 7.5%

Sodium bicarbonate ($NaHCO_3$)	75.0 g
Distilled water to make	1000.0 ml

Dissolve the NaHCO$_3$ in enough water to make 1 liter. Sterilize by passing it through a Seitz filter with positive pressure. Dispense into sterile prescription bottles using sterile technique. Stopper tightly. Label and store either at room temperature or in the refrigerator.

13. Buffered Glycerin, 60.0%

a. SOLUTION A

Potassium hydroxide (KOH)	5.6 g
Distilled water to make	10.0 ml

Prepare 10 *N* KOH by dissolving the hydroxide in enough water to make 10.0 ml.

b. SOLUTION B

Potassium phosphate, monobasic (KH$_2$PO$_4$)	2.592 g
Distilled water to make	400.0 ml

Dissolve the KH$_2$PO$_4$ in enough water to make 400.0 ml. Add 600 ml of glycerin. Adjust to pH 7.4–7.6 with the 10 *N* KOH. Pour into a flask, cover with foil, and autoclave at 15 lb pressure, 45 min. Cool, and label, and refrigerate.

For use: Add an equal volume of glycerin to an equal volume of virus material. The final concentration is 50.0% virus and 30.0% glycerin.

14. Neutral Red, 1:1000 [58]

Neutral red dye	0.1 g
Distilled water to make	100.0 ml

Place powder in a volumetric flask. Slowly add water and mix well until the powder dissolves. Make up to a final volume of 100 ml with water. Pour into a flask, cover with foil. Autoclave at 15 pounds pressure, 15 minutes. Cool. Insert a sterile rubber stopper, overwrap with foil. Store at room temperature.

15. Phenol Red, 0.5%

Phenol red dye (water soluble)	5.0 g
Sodium hydroxide (NaOH), 0.1 *N*	amount required
Distilled water to make	1000.0 ml

Add 0.1 *N* NaOH, drop by drop, to the phenol red powder until the powder dissolves. Add water to the 1-liter mark. Pour into prescription bot-

tles, cover with foil. Autoclave at 15 lb pressure, 15 min. Cool. Insert a sterile rubber stopper and overwrap with foil. Label and store at room temperature or in the refrigerator.

16. Dextrose-Gelatin-Veronal [5]

a. SOLUTION A

Veronal (barbital) $(C_8H_{12}N_2O_3)$	0.58 g
Sodium veronal (sodium barbital)	
$(C_8H_{11}N_2NaO_3)$	0.38
Gelatin	0.60
Distilled water to make	250.0 ml

Dissolve the ingredients in enough distilled water to make 250 ml, using heat.

b. SOLUTION B

Calcium chloride $(CaCl_2)$	0.02 g
Magnesium sulfate $(MgSO_4 \cdot 7H_2O)$	0.12
Sodium chloride (NaCl)	8.50
Dextrose	10.0
Distilled water to make	1000.0 ml

Dissolve the ingredients in water in a 1-liter volumetric flask. Add the 250 ml of Solution A to the flask. Bring the volume to 1 liter with water. Final pH should be 7.0–7.6. Transfer to a 2-liter flask; cover with foil. Autoclave at 10 lb pressure, 10 min. Cool. Insert a sterile rubber stopper, overwrap with foil, label, and refrigerate.

17. Acid-Citrate-Dextrose (ACD) [5]

Sodium citrate $(Na_3C_6H_5O_7 \cdot 2H_2O)$	11.26 g
Citric acid $(H_3C_6H_5O_7 \cdot H_2O)$	4.0
Dextrose	11.0
Distilled water to make	500.0 ml

Dissolve the ingredients in enough water to make 500 ml. Pour into a 1-liter flask; cover with foil. Autoclave at 10 lb pressure, 10 min. Cool. Insert a sterile rubber stopper and overwrap with foil. Label and refrigerate.

18. Veronal Buffer

a. SOLUTION A

Sodium chloride (NaCl)	83.80 g
Sodium bicarbonate $(NaHCO_3)$	2.52

Sodium veronal (sodium barbital)
($C_8H_{11}N_2NaO_3$)   3.00
Distilled water to make   1000.00 ml

Dissolve the ingredients in enough water to make 1 liter.

b. SOLUTION B

Sodium barbital   4.6 g
Magnesium chloride ($MgCl_2 \cdot 6H_2O$)   1.0
Calcium chloride ($CaCl_2 \cdot H_2O$)   0.2
Distilled water to make   500.0 ml

Dissolve the ingredients in enough hot water to make 500 ml. Add Solution A to Solution B. Bring to a volume of 2000 ml with water (this makes a 5X concentration). Sterilize by Seitz filtration and collect in sterile prescription bottles. Insert sterile stoppers and overwrap with foil. Label and refrigerate.

c. WORKING SOLUTION

To 200 ml of the 5X concentrate, add sterile distilled water to a final volume of 1000 ml (1 : 5 dilution) on the day the buffer is to be used.

19. McIlvaine's Buffer [31]

a. CITRIC ACID STOCK SOLUTION (0.1 M)

Citric acid ($H_3C_6H_5O_7 \cdot H_2O$)   19.2 g
Methyl alcohol, 25%, to make   1000.0 ml

Dissolve the acid in enough methanol to make 1 liter. Store in an Erlenmeyer flask.

b. DISODIUM PHOSPHATE STOCK SOLUTION (0.2 M)

Sodium phosphate, dibasic (Na_2HPO_4)   28.4 g
Methyl alcohol, 25%, to make   1000.0 ml

Dissolve the Na_2HPO_4 in enough methanol to make 1 liter. Store in an Erlenmeyer flask.

c. WORKING SOLUTION

Prepare a working solution of the required pH by adding the proper amounts of sodium phosphate and citric acid solutions (below) to a 200-ml volumetric flask.

Final pH	x ml 0.2 M Na$_2$HPO$_4$	y ml 0.1 M citric acid	Final pH	x ml 0.2 M Na$_2$HPO$_4$	y ml 0.1 M citric acid
2.1	4.0	196	5.2	107.2	92.8
2.4	12.4	187.6	5.4	111.5	88.6
2.6	21.8	178.2	5.6	116.0	84.0
2.8	31.7	168.3	5.8	120.9	79.1
3.0	41.1	158.9	6.0	126.3	73.7
3.2	49.4	150.6	6.2	132.2	67.8
3.4	57.0	143.0	6.4	138.5	61.5
3.6	64.4	135.6	6.6	145.5	54.5
3.8	71.0	129.0	6.8	154.5	45.5
4.0	77.1	122.9	7.0	164.7	35.3
4.2	82.8	117.2	7.2	173.9	26.1
4.4	88.2	111.8	7.4	181.7	18.3
4.6	93.5	106.5	7.6	187.3	12.7
4.8	98.6	101.4	7.8	191.5	8.5
5.0	103.0	97.4	8.0	194.5	5.5

20. Sorensen's Phosphate Buffer [32]

a. 0.15 M MONOBASIC POTASSIUM PHOSPHATE

Potassium phosphate, monobasic (KH$_2$PO$_4$)	9.08 g
Distilled water to make	1000.0 ml

Dissolve the KH$_2$PO$_4$ in enough water to make 1 liter. Place in an Erlenmeyer flask and cover with foil. Autoclave at 15 lb pressure, 15 min.

b. 0.15 M DIBASIC SODIUM PHOSPHATE

Sodium phosphate, dibasic (Na$_2$HPO$_4$)	23.87 g
Distilled water to make	1000.0 ml

Dissolve the Na$_2$HPO$_4$ in enough distilled water to make 1 liter. Place in an Erlenmeyer flask and cover with foil. Autoclave at 15 lb pressure, 15 min.

c. SALINE, 0.765%

Sodium chloride (NaCl)	7.65 g
Distilled water to make	900.00 ml

Dissolve the NaCl in enough water to make 900 ml.

d. WORKING SOLUTION

Prepare a working solution of the proper pH by mixing the phosphate solutions in the quantities given below.

pH	0.15 M Na$_2$HPO$_4$	0.15 M KH$_2$PO$_4$
5.288	0.25	9.75
5.589	0.5	9.5
5.906	1.0	9.0
6.239	2.0	8.0
6.468	3.0	7.0
6.643	4.0	6.0
6.813	5.0	5.0
6.979	6.0	4.0
7.168	7.0	3.0
7.381	8.0	2.0
7.731	9.0	1.0
8.043	9.5	0.5

Add 100 ml of the buffer mixture to 900 ml of saline.

21. Antibiotics

You may have occasion to make individual antibiotic stock solutions for addition to media. In the preparation of culture media, it is common practice to prepare a stock combination whereby several antibiotics can be added at once by a single addition of a specified volume.

a. INDIVIDUAL STOCK SOLUTIONS

(1) *Penicillin—1,000,000 units per bottle*
Add 5 ml diluent to obtain 200,000 units per milliliter (U per ml); i.e.,

$$\frac{1,000,000}{5} = 200,000$$

Add 10 ml diluent to obtain 100,000 U per ml; i.e.,

$$\frac{1,000,000}{10} = 100,000$$

(2) *Streptomycin—1-g bottle*
Add 5 ml diluent to obtain 0.2 g per ml; i.e.,

$$\frac{1}{5} = 0.2$$

Add 10 ml diluent to obtain 0.1 g per ml; i.e.,

$$\frac{1}{10} = 0.1$$

(3) *Mycostatin—500,000 Units per bottle*
Add 5 ml diluent to obtain 100,000 U per ml; i.e.,

$$\frac{500,000}{5} = 100,000$$

Add 10 ml diluent to obtain 50,000 U per ml; i.e.,

$$\frac{500,000}{10} = 50,000$$

b. DETERMINATION OF STOCK REQUIREMENTS

Decide on the final concentration of antibiotic desired in the material to be treated. To determine the amount of stock required to attain that concentration,

$$\text{Milliliters of medium (cell cultures, inoculum, etc.) to be treated} \times \frac{\text{final concentration}}{\text{of antibiotic desired}} = \frac{\text{amount of stock}}{\text{required}}$$

EXAMPLE

You have 500 ml of cell cultures. You desire 200 units of penicillin per milliliter. The stock penicillin solution contains 100,000 U per ml; therefore,

$$500 \times 200 = X \text{ ml of penicillin (100,000 U) needed}$$
$$100,000 = 100,000X$$
$$X = 1$$

Therefore, 1 ml of the 100,000 U per ml stock solution should be added to the 500 ml of cell cultures.

EXAMPLE

You wish to treat 1000 ml of medium such that the final concentration of streptomycin is 0.0001 g per ml. The stock streptomycin solution is 0.1 g per ml; therefore,

$$1000 \times 0.0001 = 0.1X$$
$$0.1 = 0.1X$$
$$X = 1$$

Therefore, 1 ml of stock should be added to 1000 ml of medium.

c. COMBINATION STOCK SOLUTIONS

(1) To obtain a combination of

Penicillin	200,000 U/ml
Streptomycin	0.2 g/ml
Mycostatin	100,000 U/ml

Add 5 ml of diluent to the bottle of penicillin. Mix. Transfer the solution to the streptomycin bottle. Mix. Transfer the solution to the mycostatin bottle. Mix. Store in the freezer.

(2) To obtain a combination of

Penicillin	100,000 U/ml
Streptomycin	0.1 g/ml
Mycostatin	50,000 U/ml

add 10 ml of diluent to the penicillin bottle. Mix. Transfer serially to the next two bottles. Freeze.

d. FOR MEDIA USE

Add 0.1 ml of the previous combination to each 100 ml of medium. This will give a final concentration of 100 units of penicillin, 0.01 g streptomycin, and 50 units of mycostatin per milliliter of medium.

22. Disinfectants

a. IODINE, WEAK TINCTURE

Iodine	2.0 g
Sodium iodide (NaI)	2.0 to 4.0
Ethyl alcohol, 95%, to make	100 ml

Dissolve 2 to 4 g of NaI in 5 ml of distilled water. Add the iodine. Dilute to 100 ml with 95% ethyl alcohol.

b. IODINE, STRONG TINCTURE

Iodine	7.0 g
Sodium iodide (NaI)	5.0
Ethyl alcohol to make	100.0 ml

Dissolve the NaI in 5.0 ml of distilled water. Add the iodine. Dilute to 100 ml with ethyl alcohol.

C.   COMMERCIALLY AVAILABLE

See Appendix B and Section G, this chapter, for a list and a discussion of commercially available disinfectants used for glassware and equipment in virologic studies.

23. Serum, Commercially Available

There are many sera which are required in the growth and maintenance media of various kinds. It is more convenient to purchase these from commercial firms. These sera are available as fetal or adult samples. The quantities you may order will vary with the amounts you normally use. It is more practical to purchase the larger quantities. These should be transferred, using sterile techniques, to smaller sterile containers. The authors have found that it is best to store sera in the ultralow temperature freezer. The keeping quality is much improved over storage in a cold room.

24. Rappaport Stain [38]

Crystal violet	0.1 g
Citric acid	1.92
Distilled water to make	100.00 ml

Dissolve the citric acid in enough water to make 75.0 ml. Add the crystal violet. Dilute to 100 ml with water. Pour into a bottle and stopper tightly. Label and store at room temperature.

25. Neutral Buffered Formalin Fixative

Formalin (40% formaldehyde in water)	100.0 ml
Sodium phosphate, dibasic, (Na_2HPO_4)	7.83 g
Sodium phosphate, monobasic, (NaH_2PO_4)	4.21
Sodium chloride (NaCl), 0.85%, to make	1000.0 ml

Neutralize the formalin by adding an excess of magnesium carbonate ($MgCO_3$). Dilute to 1 liter with 0.85% sodium chloride.

26. Carnoy's Fixative

Absolute ethyl alcohol	300.0 ml
Chloroform	150.0
Glacial acetic acid	50.0

Mix the alcohol and the chloroform. Add the acetic acid with care. Store in a tightly stoppered bottle.

27. Hematoxylin Alum

Hematoxylin	2.5 g
Ammonium alum [$(NH_4)_2SO_4Al_2(SO_4)_3 \cdot 2H_2O$]	25.0
Sodium iodate ($NaIO_3$)	0.5
Glycerol	150.0 ml
Glacial acetic acid	10.0

Dissolve the solids in 350 ml of distilled water. Add the glycerol. Carefully add the acetic acid. If a residue remains, filter.

28. Bicarbonate Wash, 1.0%

Sodium bicarbonate ($NaHCO_3$)	5.0 g
Distilled water to make	500.0 ml

Dissolve the $NaHCO_3$ is enough water to make 500 ml.

29. Eosin Counterstain

Eosin Y	2.5 g
Distilled water to make	500.0 ml

Dissolve the Eosin Y in enough water to make 500 ml.

30. Schiff Reagent

Basic fuchsin	4.0 g
Potassium metabisulfite ($K_2S_2O_5$)	7.6
Hydrochloric acid (HCl), 0.15 N	400.0 ml
Activated charcoal up to	1000.0 mg

Dissolve the basic fuchsin and the metabisulfite in enough acid to make 400 ml. Shake at intervals for 2 hr until the dye has become fuchsin-sulfurous acid. The solution should be strawcolored. Add 500 mg of fresh activated animal charcoal to decolorize. Filter. Decolorize again with remaining charcoal if necessary. Store in small dark glass bottles, in a refrigerator. Stored this way, the reagent has been known to be useful for as long as several months. Bring to room temperature before use. Do not reuse any of this reagent.

31. Sulfite Wash

Sodium metabisulfite ($Na_2S_2O_5$), 10%	60.0 ml
1 N HCl	60.0
Distilled water to make	1000.0 ml

Dissolve the metabisulfite in sufficient water to make 1000 ml. Add the HCl.

32. Fast Green Counterstain

Fast green	0.03 g
Ethyl alcohol, 95%, to make	300.0 ml

Dissolve the dye in enough alcohol to make 300 ml.

33. Control Enzymes [34]

a. PEPSIN

Pepsin, X crystallized	0.02 g
0.02 M HCl to make	100.0 ml

Dissolve the pepsin in enough HCl to make 100 ml.

b. RNASE

RNase, 5X crystallized	0.05 g
Distilled water, pH 7, to make	100.0 ml

Dissolve the RNase in enough distilled water to make 100 ml.

c. DNASE

DNase, X crystallized	0.01 g
Veronal buffer, 0.025 M, containing	
0.003 M $MgSO_4$ to make	100.00 ml

Dissolve the DNase in enough veronal buffer to make 100 ml.

34. Inactivated Control Enzymes

Prepare enzymes as indicated above. Inactivate by boiling (100°C) for 1 hr or by autoclaving at 15 lb pressure for 10 min.

35. Acridine Orange Fluorochrome Stain

Acridine orange	0.03 g
McIlvaine's buffer to make	300.00 ml

Dissolve the acridine orange in enough McIlvaine's buffer to make 300 ml.

36. Phosphate Buffered Saline (PBS), 10X [13]

Concentrated (10X) PBS is prepared in three separate solutions, in order to avoid precipitation of salts.

a.   SOLUTION A

Sodium chloride (NaCl)	80.0 g
Potassium chloride (KCl)	2.0
Sodium phosphate, dibasic (Na_2HPO_4)	11.5
Potassium phosphate, monobasic (KH_2PO_4)	2.0
Distilled water to make	1000.0 ml

Add the ingredients, in the order listed, to 500 ml of water in a graduate cylinder. Stir after the addition of each chemical. Make the volume to 1 liter with water. Pour the solution into a 2-liter Erlenmeyer flask and cover with a double fold of aluminum foil. Label.

b.   SOLUTION B

Calcium chloride:	
as $CaCl_2 \cdot 2H_2O$	1.3 g
or as $CaCl_2$	(0.5 g)
Distilled water to make	100.0 ml

Dissolve the calcium chloride in enough water to make 100 ml. Pour into a 250-ml Erlenmeyer flask and cover with a double fold of aluminum foil. Label.

c.   SOLUTION C

Magnesium chloride ($MgCl_2 \cdot 6H_2O$)	1.0 g
Distilled water to make	100.0 ml

Dissolve the magnesium chloride in enough water to make 100 ml. Pour into an Erlenmeyer flask and cover with a double fold of aluminum foil. Label.

Autoclave the three solutions at 15 lb pressure, 15 min. Cool. With sterile technique, add sterile rubber stoppers.

(1) Flame the aluminum foil cover and loosen, but do not remove it.

(2) Remove a sterile rubber stopper from its container, lift the cap from the flask, and quickly put the stopper in place.

(3) Replace the foil cap over the stopper. Refrigerate the three solutions.

37. Phosphate Buffered Saline, Working Solution

A detailed description of the procedure for preparing PBS working solution follows. Once you have prepared PBS, you should be relatively adept in the art of flaming flasks, graduate cylinders, stoppers, and foil covers before and after use, seemingly all at once. This technique should be learned thoroughly and *must* be followed when preparing all the media.

    a. Formula

Solution A, sterile	400 ml
Distilled water, sterile	4500
Solution C, sterile	50
mix	
Solution B, sterile	50

    b. Wash the work bench with disinfectant.

    c. Flame the aluminum cover of a 1-liter graduate cylinder and loosen, but do not remove, the foil.

    d. Flame the aluminum cover of the 10X Solution A and loosen the cover.

    e. Holding the flask of Solution A in the right hand, remove the foil and place it upside down on the bench top. Remove the rubber stopper and place it bottom down in the foil cover. Flame the mouth of the flask.

    f. Remove the cover of the graduate cylinder with the left hand to the degree that liquid can be poured easily into the cylinder, but do *not* remove the cover entirely. Pour 400 ml of Solution A into the graduate cylinder. Drop the cover back on the cylinder.

    g. Flame the mouth of the Solution A flask; flame the stopper by passing it quickly through the Bunsen flame and place it tightly back into the flask. Flame the aluminum cover and place it over the stopper.

    h. Flame and loosen the cover of a 5-liter flask.

    i. Holding the graduate cylinder in the right hand, remove the cover with the left hand and hold it between the index and middle fingers. Flame the mouth of the cylinder.

    j. Remove the cover of the 5-liter flask enough to allow the liquid to be added. Pour the 400 ml of Solution A into the flask. Replace the foil cover of the flask.

    k. Flame the mouth of the cylinder, flame the foil cap, and recover the cylinder.

    l. Continue the method of flaming the cylinders, flasks, and foil covers through the subsequent procedures.

    m. Add 4500 ml of sterile distilled water to Solution A in the 5-liter flask. Mix thoroughly.

n. Add 50 ml of Solution C to the 5-liter flask. Mix thoroughly.
o. Add 50 ml of Solution B to the flask with constant manual swirling. Addition of the solutions in this order will prevent precipitation of the salts.
p. With sterile technique, fill 16-oz prescription bottles. If you do not have steady hands, you will find that a sterile glass funnel will be helpful. Stopper each bottle with sterile black rubber stoppers and cover with an aluminum foil cap.
q. Label each bottle. Refrigerate.

38. Trypsin, 0.25% [42]

Trypsin is commercially available as a sterile, 10X concentrated solution. Where money is no object, it is recommended that this be used as follows:

10X Trypsin solution, sterile	100.0 ml
PBS, sterile to make	1000.0 ml

With sterile technique, dilute the 10X trypsin solution with enough sterile PBS to make 1 liter. Dispense into sterile bottles, stopper, and cover with foil. Label and refrigerate.

Powdered trypsin may be prepared as follows:

Trypsin powder (Difco, 1:250)	6.25 g
PBS, sterile to make	2500.00 ml

Add the trypsin to PBS and dilute to 1 liter. Allow the solution to clarify at room temperature. Seitz filter it under positive pressure. Distribute to sterile bottles, stopper, and cover with foil. Label and refrigerate.

39. Hanks' 10X Medium, Sterile [22, 23, 60]

a. SOLUTION A

Sodium chloride (NaCl)	160 g
Potassium chloride (KCl)	8
Magnesium sulfate (MgSO$_4 \cdot$7H$_2$O)	4
Distilled water to make	800 ml

Dissolve the ingredients in enough water to make 800 ml.

b. SOLUTION B

Calcium chloride:	
as CaCl$_2$	2.8 g
or as CaCl$_2 \cdot$2H$_2$O	(4.1)
Distilled water to make	200.0 ml

Add the calcium chloride to enough water to make 200 ml. Add Solution A to Solution B and pour into a 2-liter flask. Cover with foil.

c. SOLUTION C

Potassium phosphate, monobasic (KH_2PO_4)	1.2 g
Sodium phosphate, dibasic (Na_2HPO_4)	1.2
Dextrose	20.0
Phenol red, water soluble	0.4
Distilled water to make	1000.0 ml

Dissolve the ingredients in enough water to make 1 liter. Pour into a 2-liter flask and cover with foil. Autoclave Solution A-B and Solution C at 10 lb pressure for 10 min. Cool. With sterile technique, combine the two solutions. Insert a sterile rubber stopper and overwrap with foil. Label and refrigerate.

40. Hanks' Growth Medium

Hanks' 10X medium, sterile	500 ml
Distilled water, sterile	4000
Lactalbumin hydrolysate, 5%, sterile	500

With sterile technique, add the ingredients to a 5-liter flask. Add 1 million units penicillin, 1 g streptomycin, and 500,000 units mycostatin. Mix thoroughly. Then add

Sodium bicarbonate, 7.5%, sterile:	
if NaOH was used in the LAH	25 ml
if no NaOH was used in the LAH	35

Mix thoroughly. Dispense into sterile prescription bottles, insert stoppers, overwrap with foil, and label. Refrigerate. For use, add 10 to 20% sterile serum (type to depend on virus under study).

41. Hanks' Maintenance Medium

Hanks' 10X medium, sterile	100 ml
LAH, 5%, sterile	100
Tryptose, 5%, sterile	100
Sodium bicarbonate ($NaHCO_3$), 7.5%, sterile:	
if NaOH was used in the LAH	5
if no NaOH was used in the LAH	7
Distilled water to make	1000 ml
Powdered antibiotics, 1 bottle each	

Add the ingredients to a 1-liter flask. Mix thoroughly. Dispense into sterile prescription bottles. Insert sterile black rubber stoppers and cover with foil. Label and refrigerate. For use, add 2.0% serum of the type desired. Where viral inhibition occurs, you may eliminate serum altogether.

42. Hanks' 10X Medium Without Phenol Red

Prepare exactly as described above, omitting the dye.

43. Lactalbumin Medium With Hepes Buffer [4]

a. SOLUTION A

Sodium chloride (NaCl)	8.0 g
Potassium chloride (KCl)	0.4
Calcium chloride (CaCl$_2$)	0.14
Magnesium sulfate (MgSO$_4$)	0.2
Glucose	1.0
Phenol red	0.02
Hepes buffer	0.6
Distilled water to make	700.0 ml

Dissolve the ingredients in enough distilled water to make 700 ml. Then add Solution B.

b. SOLUTION B

Lactalbumin hydrolysate (LAH)	100 ml
Calf serum	50
Sodium bicarbonate (NaHCO$_3$)	14
Antibiotic mixture	1

Adjust the pH to 7.3–7.4 with 10% sodium hydroxide solution (about 3.0 ml). Bring to a final volume of 1000 ml with distilled water. Filter sterilize through a sterile 0.22 μm membrane filter into a sterile receiving flask. With sterile technique, dispense into prescription bottles, stopper, and overwrap with foil. Label and refrigerate.

44. Earle's Balanced Salt Solution, 10X [14]

a. SOLUTION A

Magnesium chloride (MgCl$_2$·6H$_2$O)	203 g
Calcium chloride (CaCl$_2$)	4
Distilled water to make	800 ml

Dissolve the ingredients in enough water to make 800 ml.

b. SOLUTION B

Sodium chloride (NaCl)	272 g
Potassium chloride (KCl)	6
Magnesium sulfate ($MgSO_4 \cdot 7H_2O$)	8
Sodium phosphate, monobasic ($NaH_2PO_4 \cdot H_2O$)	5
Glucose	40
Distilled water to make	3200 ml

Dissolve the ingredients in enough distilled water to make 3200 ml. Autoclave both solutions at 15 lb pressure, 15 min. Cool. Slowly pour Solution A into Solution B (*not* vice versa) while slowly swirling the flask. Stopper and overwrap with foil. Label and refrigerate.

45. Earle's Balanced Salt Solution: Working Solution

Earle's 10X solution, sterile	100 ml
Distilled water, sterile	882
Phenol red, 0.5%, sterile	4
Sodium bicarbonate ($NaHCO_3$), 7.5%, sterile	14

With sterile technique, pour the ingredients into a 1-liter flask. Stopper and cover with foil. Label and refrigerate.

K.   GENERAL EQUIPMENT

There are several pieces of additional equipment which are needed in any virus laboratory for tissue culture, egg propagation of viruses, histologic examination of tissue, etc. These include racks to hold tubes containing cell monolayers, canisters to hold tubes of cells, etc. These special pieces are available commercially and can be expensive. With a little ingenuity, you can make very serviceable substitute equipment.

1. Canisters for Test Tube Cultures

a. STAINLESS STEEL

Stainless steel canisters, approximately 8 × 8 × 12 in. can be purchased. They are strong, durable, and easy to clean. They can be expensive to buy or to have made.

b. CARDBOARD

Test tubes and other pieces of small equipment come in boxes approximately 8 × 8 × 12 in. and these can be put to use very easily:

(1) Remove the cover or the fold-in top of the box.

(2) Turn the box so that it is resting on its 12-in. side.

(3) Cut three thin pieces of cardboard (file cards or the test tube box separators are satisfactory for this) about 2 in. wide and as long as the box.

(4) Place these three strips of cardboard on top of each other under the bottom edge of the box [Fig. 9-27(a)].

(5) Cut a length of wide masking tape 12 in. long. Lay the tape over the front edge of the box and fold it over the strips and down under the box. You now have a box with the bottom front edge adjusted to an approximately 5° lift [Fig. 9-27(b)].

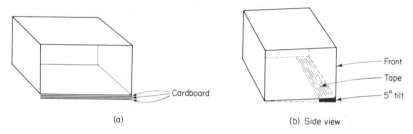

(a) (b) Side view

Fig. 9-27. Test tube box construction.

2. Test Tube Culture Racks

a. STAINLESS STEEL

Stainless steel tissue culture racks are commercially available (see Appendix B). They must be fixed to prevent the tubes from falling out and to minimize rotation during handling (i.e., the monolayer might come to rest above the level of the medium and will die).

(1) Cut a strip of cardboard the length of the back (bottom) part of the rack (i.e., the part which has no holes).

(2) Starting at one end of the cardboard strip, wind wide masking tape, sticky-side out, over it in a continuous overlapping strip (Fig. 9-28). You now have a length of cardboard which has sticky-surface tape on both its front and its back.

(3) Very carefully lower this strip down between the middle (which has holes) and the bottom (no holes) shelves (Fig. 9-29). Although Fig. 9-29 shows a rack of wood frame with a masonite tube grid, the insertion of a tape-wound backer board is similar; with stainless steel racks, the backer board is simply pressed against the tube side of the bottom; with the wooden rack shown, the backer board is dropped into the slot provided.

(4) Press the strip against the bottom (back) of the rack so that the

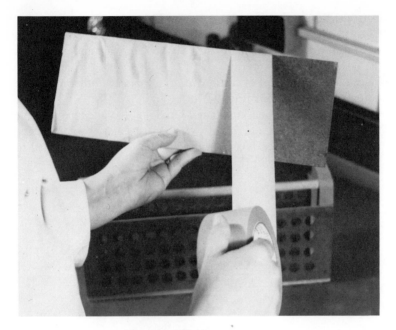

Fig. 9-28. Masking tape.

Fig. 9-29. Backer Board insertion.

strip attaches thereto. If you now put a test tube into the rack, it will go through the holes of the top and middle shelves to rest in contact with the tape.

b. WOOD

If stainless steel racks are at a premium or are too expensive to be purchased, wooden racks can be made by a carpenter or home hobbyist as follows.

(1) Cut two pieces of pressboard, 5 × 15 in. and, in each, cut out four rows of 15 holes which are $\frac{11}{16}$ in. in diameter (Fig. 9-30).

(2) Cut two pieces of $\frac{3}{4}$-in. pine into $2\frac{7}{8}$ × $6\frac{1}{2}$ in. blocks (front end pieces).

(3) Cut two pieces of $\frac{3}{4}$-in. pine into $1\frac{1}{8}$ × 5 in. blocks (rear end pieces) and cut a $\frac{1}{4}$-in. deep groove in the 5 in. length of the two blocks (Fig. 9-31). The groove should be wide enough to allow insertion of the depth of a piece of pressboard.

(4) Cut one more piece of pressboard, 5 × $13\frac{7}{8}$ in. (back).

(5) Cut a 15-in. length of $\frac{1}{2}$-in. diameter dowel for the handle.

(6) Assemble the pieces as illustrated (Fig. 9-31).

(7) Overwrap the back piece of pressboard with masking tape. Slip the back piece into the slotted grooves of the rear end pieces.

(8) Cut a length of any kind of paper and stick it to the outer exposed sticky bottom panel.

(9) Cut a 15 × $\frac{3}{4}$ in. strip of $\frac{1}{4}$-in. lath and nail this to the bottom front edge of the rack so that the rack has a 5° backward tilt when it rests on the bench top.

3. Micro-Slide Carry-All

It is often necessary to study infected monolayers which have been stained with suitable dye (Chapter 7). The cell sheets, on 8 × 22 mm cover slips, are far too small to be processed in a conventional slide carrier. It is also

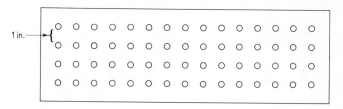

Fig. 9-30. Construction drawing for tube grid.

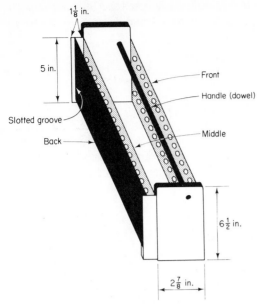

Fig. 9-31. Construction diagram for wooden test tube rack.

difficult to process them manually with forceps since invariably the cover slips are dropped, broken, or chipped at some stage in the staining process.

In order to solve this problem, Dr. Theodore Chomiak (Department of Pathobiology, University of Connecticut, Storrs, Conn.) designed an excellent micro-slide carrier, constructed as follows.

    a. Remove the two ridged sides of a conventional metal slide carrier (Fig. 9-32).

    b. Take a conventional metal slide carrier which has not been dis-

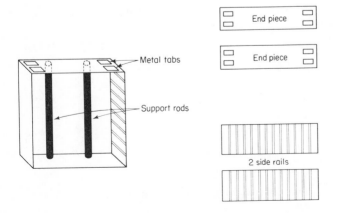

Fig. 9-32. Micro carry-all construction.

mantled. Place the loose side rails (step a) into the carrier with the notches facing the sides of the carrier. Rest the rails on the support rods (Fig. 9-33).

c. Solder the rails in place at both ends. You now have a carrier with double supports for 8 × 22 mm cover slips (Fig. 9-33). Attach a standard handle down the center.

Another method is to make a single carrier by soldering the two loose side rails (step a) together, at an angle. Then, solder this between the two end pieces and attach a standard handle (Fig. 9-34).

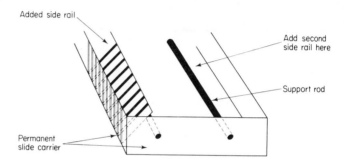

Fig. 9-33. Construction for double micro-slide carry-all.

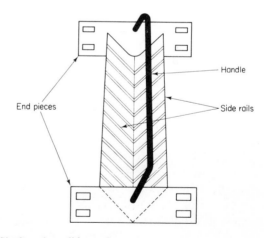

Fig. 9-34. Single- micro-slide carrier.

4. Egg Punch

a. Stainless steel egg punches may be purchased commercially. Essentially, they are a cylindrical stainless steel barrel with a short pointed probe on one end [Fig. 9-35(a)].

b. Some workers use a dental or hobbyist's drill with a burr attachment. The drill may prove to be weighty for some to handle, and you must have a light touch on the surface of the egg to assure that the burr does not penetrate the shell membrane. The drill also becomes hot with prolonged use.

c. A large 18-gauge needle, with the tip blunted, pushed through a small cork may be used [Fig. 9-35(b)]. The cork then serves as a handle.

d. You may also use a dissecting probe. Use tin snips to cut the needle off close to the wooden handle [Fig. 9-35(c)]. File the needle until it is smooth.

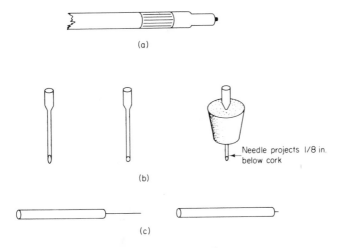

Fig. 9-35. Egg punches.

5. Candling Lamp

There are many ways to construct a candling lamp. The important point to realize is that a concentrated beam of light is needed. One method is to take an old lamp cord plus the attached socket, a bulb, and a can or piece of sheet metal. Choose a small can whose diameter is smaller than the average breadth of an egg. Remove both ends of the can and assemble into a permanent unit (Fig. 9-36). A larger can is best prepared by removing only the bottom. Cut a hole in the side of the can large enough to accommodate the breadth of an egg. Assemble as above.

6. Flat Trays

Flat trays with $1\frac{1}{2}$ to 2 in. deep sides are convenient for holding and transporting agar overlays and Petri dish cultures. Commercial ones of

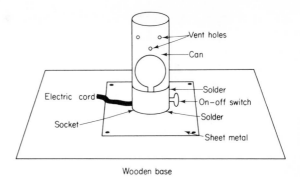

Fig. 9-36. Candling lamp.

stainless steel are excellent since they have smooth corners and can be cleaned easily. Flat wooden trays can be made inexpensively in any size desired; $2 \times 10 \times 15$ in. have been found convenient.

    a. Cut two pieces of 2 in. wide lath, 10 in. long.

    b. Cut two pieces of 2 in. wide lath, 15 in. long.

    c. Nail together at the corners. Cut a piece of masonite board as wide as the outer edge of the frame, but about 2 in. shorter than the length of the frame (this will allow for circulation of heat). Nail to the bottom of the frame (Fig. 9-37).

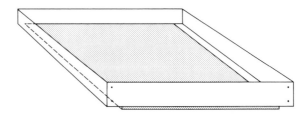

Fig. 9-37. Wooden tray construction.

7. Freezer Boxes

Freezer boxes made of corrugated board may be purchased commercially (see list of suppliers, p. 273). Any strong material can be converted into a box of the right dimensions. The bottom should measure $15\frac{5}{8} \times 6\frac{1}{2} \times 5\frac{3}{8}$ in. The cover should measure $16\frac{1}{4} \times 7 \times 2\frac{1}{4}$ in. (Fig. 9-38).

An internal compartmented section of heavy cardboard strips is made as follows.

    a. Cut three pieces $15\frac{3}{8} \times 4\frac{1}{2}$ in. and make 10 diagonal slashes as illustrated in Fig. 9-39(a).

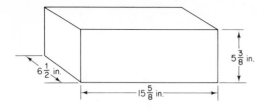

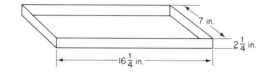

Fig. 9-38. Freezer box.

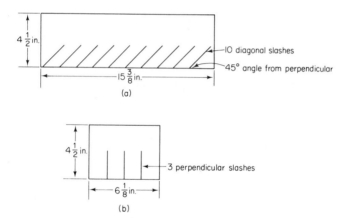

Fig. 9-39. Freezer box, internal compartments.

b. Cut 10 pieces $6\frac{1}{8} \times 4\frac{1}{2}$ in. and make three perpendicular slashes as illustrated in Fig. 9-39(b).

c. Put the two units together in the area of the slashes, the 10 short pieces across (width of the box) and the long pieces lengthwise (Fig. 9-40). The final effect is that of an egg crate, slightly slanted.

d. Mark the inside of the box with a brush pen (Fig. 9-41); put the numbers on the inside top edge of the long dimension and the letters on the inside top edge of the short dimension.

e. Mark the outside of the box (the same end marked inside as

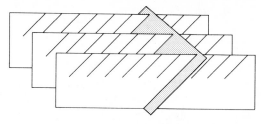

Partitions turned upside down

Fig. 9-40. Assembling internal compartments.

10	9	8	7	6	5	4	3	2	1 A
									B
									C
									D

Top view

Fig. 9-41. Marking internal compartments.

A, B, C, D) as the *top*. When the box is placed in the freezer, it will be standing on end (Fig. 9-42). Due to the slant arrangement of the internal compartments, the tubes will be tilted, with the lips upward. Materials may be stored in a liquid state and they can be frozen without fear of spillage. If several individuals use the freezer, each box should be marked with individual numbers. This makes for rapid identification when the freezer is opened.

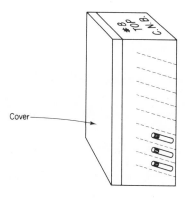

Cover

Fig. 9-42. Assembled box.

8. Freezer Record Card

An accurate record of frozen-stored virus or other materials should be kept. Experience suggests that 4 × 6 in. white cards are excellent, although pages in a book may be used. If the entries under "Description" (Fig. 9-43) are kept in pencil, while other permanent headings are in ink or typed, the card may be used repeatedly as materials are stored, used, or discarded. Whether cards or a record book are used, their placement near the freezers concerned will encourage the keeping of accurate records as the freezer contents are used and replaced.

Freezer No. 3	Box No. 2	Owner: ROVOZZO
Column	Compartment	Description
A	1	ECBO # 48, 3rd calf kid. pass.
	2	ECBO # 93, 4th calf kid. pass.
	3	
	4	
	5	
B	1	
	2	

Fig. 9-43. Freezer record card.

APPENDIX

A. Methods of Calculation

A. DETERMINATION OF METRIC EQUIVALENTS

In one
of these, there are _____ of these ——→

	m (meter)	cm (centi- meter)	mm (milli- meter)	μ (micron) or μm (micrometer)	mμ (millimicron) or nm (nanometer)	Å (angstrom)
m	1	100	1000	1,000,000	1,000,000,000	10,000,000,000
cm (10^{-2})		1	10	10,000	10,000,000	100,000,000
mm (10^{-3})			1	1000	1,000,000	10,000,000
μ (10^{-6})				1	1000	10,000
mμ (10^{-9})					1	10
Å (10^{-10})						1

EXAMPLE

To determine the metric equivalent of any unit in the vertical column read positively from left to right
i.e., 1 m is equivalent to 1000 mm.
i.e., 1 mm is equivalent to 10,000,000 Å.
To determine the metric equivalent of any unit in the horizontal row, read negatively from right to left
i.e., 1 nm is equivalent to 1 millionth ($\frac{1}{1,000,000}$) of a mm.
i.e., 1 Å is $\frac{1}{10,000,000}$ mm.
i.e., 1 μ is $\frac{1}{1000}$ mm.

B. DETERMINATION OF DILUTIONS

There are three ways to prepare a dilution. Suppose you are asked to prepare a 1 : 10 dilution. Essentially, this means you must prepare 1 part of solute in a *total* of 10 parts of solvent. What are the ways this may be done?

1. Method 1: Weight:Weight

For most purposes, one can assume that 1 ml of liquid (solvents or solutes) weighs the same as 1 ml of water (1 g). For precision, however, all

259

liquids should be weighed if weight : weight (w : w) combinations are specified. To prepare a 1 : 10 (w : w) dilution, combine 1 g of solute with 9 g of
solvent to give a total of 10 g.

2. Method 2: Weight : Volume

This is probably the most common method used. To prepare a 1 : 10
dilution by weight : volume (w : v), weigh out 1 g of solute, place it in a graduate cylinder, and fill with solvent to the 10-ml mark. One then has 1 g of
solute in a total volume of 10 ml.

One important point should be made here. When the amount of solute is
small (i.e., a dilute solution), the volume will not change to any perceptible
degree but the weight of solution will definitely change. When the amount
of solute to be used is large (i.e., a more concentrated solution), the volume
will be seen to increase (or decrease) considerably. One should make it a
practice, therefore, to dissolve the required amount of solute in enough
solvent to give the *total* volume specified.

3. Method 3: Volume : Volume

To prepare a 1 : 10 dilution by volume : volume (v : v), add 1.0 ml of
liquid to 9.0 ml of liquid for a total volume of 10.0 ml.

The important fact to remember is that the solute : solvent ratio is 1 part
to a total of 10 parts. The exact meaning of *part* in each case should be
specified so that others understand your dilution method and can duplicate
your work.

4. Per Cent Solutions

In the same way, a *per cent* solution means X many parts per 100 parts.
Each of the solutions discussed above could be designated in terms of per
cent (since 1 part in 10 is the same as 10 parts in 100).

    a. 10% (w : w) is prepared by dissolving 1 g of solute in 9 g of
       solvent to give a total of 10 g.

    b. 10% (w : v) is prepared by dissolving 1 g of solute in enough
       diluent to give a total volume of 10 ml.

    c. 10% (v : v) is prepared by combining 1 ml of solute A with 9 ml
of solvent B to give a total volume of 10 ml.

5. Conversion Methods

One problem which is often confusing is the conversion of a solution
from one concentration or dilution to another. You can convert a more concentrated solution to a more dilute solution (i.e., a 1 : 40 dilution to a 1 : 80

dilution) but you cannot convert a more dilute solution to a more concentrated one without removing water (i.e., a 1 : 80 dilution to a 1 : 40 dilution). The important point to realize is that a 1 : 40 dilution is more concentrated than a 1 : 80 dilution, irrespective of the fact that 80 is a larger *number*. In the same way, 1 : 10 is more concentrated than 1 : 40; 1 : 320 is more concentrated than 1 : 640.

EXAMPLE

Suppose you have 10 ml of a 1 : 20 dilution and require a 1 : 60 dilution. How many times must you dilute the 1 : 20 (0.05% solution) dilution to make a 1 : 60 (0.016% solution) dilution?

$$\tfrac{1}{20} \div \tfrac{1}{60} = \tfrac{1}{20} \times \tfrac{60}{1} = 3$$

Therefore, you must dilute the original 10 ml of 1 : 20 dilution three times: 10 ml of 1 : 20 dilution + 20 ml of diluent = 30 ml of 1 : 60 dilution.

Some find it easier to convert fractions to per cent before converting concentrations.

EXAMPLE

Convert 20 ml of a 1 : 8 dilution of KCl to a 1 : 20 dilution.
    a. *Calculate the per cent concentration for each:*

$$\tfrac{1}{8} \times 100 = 12.5\%$$
$$\tfrac{1}{20} \times 100 = 5.0\%$$

    b. *Calculate the dilution factor:*

$$\frac{12.5\%}{5.0\%} = 2.5$$

    c. *Multiply the factor times the volume:*

$$2.5 \times 20 = 50.0$$

    d. Place the 20 ml of 1 : 8 KCl in a graduate cylinder and add diluent up to the 50-ml mark. You now have 50 ml of a 1 : 20 dilution of KCl.

EXAMPLE

How many grams of KCl must you dissolve in 5.0 ml of diluent to prepare a 1 : 200 dilution?

a. *Calculate the per cent concentration:*

$$\tfrac{1}{200} \times 100 = 0.5\%$$

b. 0.5% solution is equivalent to

0.5 g in 100 ml

c. *Set up a ratio:*

$$\frac{0.5 \text{ g}}{100 \text{ ml}} = \frac{X \text{ g}}{5 \text{ ml}}$$

$$100\,X = 2.5$$

$$X = 0.025 \text{ g}$$

d. Place 0.025 g KCl in a graduate cylinder. Add diluent to the 5-ml mark. You now have 5 ml of a 1 : 200 dilution.

EXAMPLE

How do you prepare a 1 : 200 dilution of KCl in a total of 4 ml, given a stock solution of 1 : 10 KCl?

a. *Calculate the per cent concentration:*

$$\tfrac{1}{10} \times 100 = 10\%$$

$$\tfrac{1}{200} \times 100 = 0.5\%$$

b. 10% solution is equivalent to

10 g in 100 ml, or 0.1 g in 1 ml

Thus, each milliliter of 1 : 10 stock solution has 0.1 g of KCl (i.e., 100 mg). Each milliliter of 1 : 200 desired solution will have 0.005 g per ml of KCl (i.e., 5 mg).

c. You need 4 ml containing 5 mg per ml; i.e., 4×5 mg $= 20$ mg from the stock.

d. $\dfrac{20 \text{ mg (required)}}{100 \text{ mg (stock)}} = \dfrac{1}{5}$

Therefore, 0.2 ml (i.e., $\tfrac{1}{5}$) will contain 5 mg KCl.

e. Add 0.2 ml stock KCl (20 mg) to 3.8 ml of diluent to obtain 4.0 ml of 0.5% solution.

EXAMPLE

In any instance where you wish to prepare a more dilute solution from a more concentrated supply, use the following system to set up the proportion: amount × concentration (solution wanted) = amount × concentration (stock). Remember, the *amount* and the *concentration* can be expressed in the proportion with any appropriate unit, i.e., g, ml, per cent, normality, molarity, etc.

    a. To prepare 50 ml of 60% ethyl alcohol, given 95% alcohol.

$$50 \text{ ml of } 60\% = X \text{ ml of } 95\%$$

$$\frac{50 \text{ ml} \times 60\%}{95\%} = X = 31.5 \text{ ml}$$

    To 31.5 ml of 95% ethanol, add 18.5 ml of water to prepare 50 ml of a 60% solution.

    b. To determine how many ml of a 0.05 *M* buffer can be made from 12 ml of a 0.2 *M* buffer.

$$X \text{ ml of } 0.05 \, M = 12 \text{ ml of } 0.2 \, M$$

$$X = \frac{12 \text{ ml} \times 0.2 \, M}{0.05 \, M}$$

$$X = 48 \text{ ml}$$

    To 12 ml of the 0.2 *M* buffer, add enough water (36 ml) to bring the total volume to the 48 ml mark.

6. Accuracy in Volumetric Methods

In all methods employed in virology, the greatest possible accuracy is required. You must use the flask, graduate cylinder, or pipet which will give you the degree of accuracy required for the particular measurement you are making.

Graduate cylinders are used where a high degree of accuracy is not required. The better cylinders, however, are manufactured with a great degree of care today. With proper use, much of the media described in this manual can be prepared in graduate cylinders.

We use a serologic pipet in preference to a transfer pipet (calibrated to drain or blow out) because it is not convenient to handle a transfer pipet during the large number of serial dilutions performed in virus work. Remember that to blow out a transfer pipet [Fig. A-1(a)] does *not* mean that you actually blow through the pipet! The small drop of fluid remaining in the tip is removed by closing the mouthpiece of the pipet with the forefinger and warming the bulb by gripping it in the palm of the left hand. Expanding air in the bulb forces the drop out of the tip. With those transfer pipets *not* graduated to the tip [Fig.A-1(b)], you must touch the tube with the tip of the pipet to deliver volumes between the graduation marks. You could not deliver fluids from these transfer pipets since they are *not* calibrated to the tip.

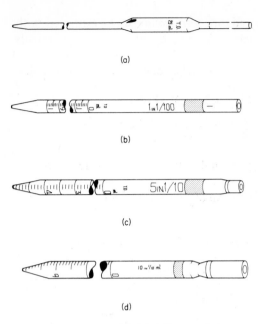

(a)

(b)

(c)

(d)

Fig. A-1. Types of pipets.

Serologic pipets which are used to deliver are usually calibrated to the tip [Fig. A-1(c)]. The volume is adjusted for blowing out (literally) the last drop. This is designated, on most brands, by two horizontal bands etched in the glass at the mouth end of the pipet.

Be sure to use serologic "cotton mouth" pipets. These are designed with a chamber at the mouthpiece end [Fig. A-1(d)] in which a cotton plug can be inserted. This type should always be used as a safeguard to you as the user and to lessen the possibility of contamination of sterile media and cultures.

Always hold a pipet between the thumb and middle finger of the right hand with the index finger (*not the thumb*) on the mouthpiece (Fig. A-2).

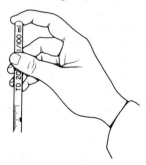

Fig. A-2. Proper holding of pipet.

With a little practice, you will find a spot on the fleshy part of the fingertip where you have excellent control over the pressure you exert on the mouthpiece. A slight decrease in pressure on the mouthpiece (you do *not* have to remove the finger) will permit a small amount of air to leak in, and the level of fluid in the pipet will drop. Increased pressure will stop the flow of liquid.

You will find that this method of holding a pipet allows for fine control and complete flexibility in handling, and it is not tiring. Never, *but never*, use the thumb for controlling a pipet—it is awkward, tiresome, and completely unprofessional!

C. DETERMINATION OF NORMALITY AND MOLARITY

A molar solution contains the molecular weight of a substance (1 gram-molecular weight) in 1 liter of solution.

EXAMPLE

You wish to prepare various molar concentrations of NaCl in water. The molecular weight of NaCl is 58.54 g.

Therefore, 1 M (molar) NaCl = 58.54 g NaCl dissolved in 1 liter of water

$$2\ M\ \text{NaCl} = 58.54\ \text{g NaCl} \times 2\ \text{or}$$
$$117.08\ \text{g NaCl per liter of water}$$

$$0.1\ M\ \text{NaCl} = 58.54\ \text{g NaCl} \times 0.1\ \text{or}$$
$$5.854\ \text{g NaCl per liter of water}$$

$$0.6\ M\ \text{NaCl} = 58.54\ \text{g NaCl} \times 0.6\ \text{or}$$
$$35.124\ \text{g NaCl per liter of water}$$

It is not necessary to prepare a liter of solution if the amount required is less.

EXAMPLE

You wish to prepare 10 ml of 1 M NaCl in water.

$$1\ M\ \text{NaCl} = 58.54\ \text{g per liter of water}$$
$$= 5.854\ \text{g per 100 ml water}$$
$$= 0.5854\ \text{g per 10 ml}$$

EXAMPLE

You wish to prepare 50 ml of 0.1 M NaCl in water.

$$1 \; M \; \text{NaCl} = 58.54 \text{ g per liter of water}$$
$$0.1 \; M \; \text{NaCl} = 5.854 \text{ g per liter}$$
$$= 0.5854 \text{ g per 100 ml}$$
$$= 0.2927 \text{ g per 50 ml}$$

If it is necessary to convert molarity to per cent, one divides the number of moles by 10.

EXAMPLE

$$1 \; M \; \text{NaCl} = 58.54 \text{ g per liter}$$
$$= 5.854 \text{ g per 100 ml}$$
$$= 5.854\%$$

EXAMPLE

$$0.1 \; M \; \text{NaCl} = 5.854 \text{ g per liter}$$
$$= 0.5854 \text{ g per 100 ml}$$
$$= 0.5854\%$$

A normal solution is a molar (gram-molecular weight) solution divided by the number of hydrogen ions present, dissolved in a liter of water.

EXAMPLE

$$1 \; M \; \text{HCl} = 36.45 \text{ g per liter}$$
$$1 \; N \; \text{HCl} = 36.45 \text{ g per liter}$$

i.e., 1 M and 1 N HCl are identical.

EXAMPLE

$$1 \; M \; \text{H}_2\text{SO}_4 = 98.0 \text{ g per liter}$$
$$1 \; N \; \text{H}_2\text{SO}_4 = \frac{98}{2} = 49 \text{ g per liter}$$

A normal solution may also be calculated by dividing the molarity by the number of potential hydrogen ions present.

EXAMPLE

$$1 \; M \; \text{NaOH} = 40 \text{ g per liter}$$
$$1 \; N \; \text{NaOH} = \frac{40}{1} = 40 \text{ g per liter}$$

since one hydrogen (H) would combine with one hydroxyl (OH) to form H_2O.

$1 \, M \, H_2CO_3 = 106$ g per liter

$1 \, N \, H_2CO_3 = \frac{106}{2} = 53$ g per liter

since two H's would combine with CO_3^{--} to form H_2CO_3.

D. HYDROUS-ANHYDROUS CHEMICAL COMPUTATIONS

There will be times when you wish to make up a solution and find that you have a hydrated form of a particular chemical you need. It is not difficult to make the conversion.

You require 10 g of dextrose $(C_6H_{12}O_6)$ but only the hydrated form $(C_6H_{12}O_6 \cdot H_2O)$ is available. How much of the hydrated form must be used to equal 10 g of anhydrous dextrose?

The molecular weight of anhydrous dextrose is 180.

The molecular weight of hydrated dextrose is 198.178.

Therefore: $10 \text{ g} \times 180 = X \text{ g} \times 198.178$

Then

$$\frac{198.178}{X} = \frac{180}{10}$$

$$180X = 198.178 \times 10$$

$$X = \frac{1981.78}{180} = 11.00$$

Therefore, use 11.00 g of hydrated dextrose to equal 10 g of anhydrous dextrose in your formula.

E. DETERMINATION OF EQUIVALENT TEMPERATURES

1. To convert degrees Celsius (C, centigrade) to degrees Fahrenheit (F):

$$\frac{9 \times {}^\circ C}{5} + 32 = {}^\circ F, \text{ or } {}^\circ F = 1.8C + 32$$

To convert 10°C to °F,

$$\frac{9 \times 10}{5} + 32 = 50{}^\circ F$$

To convert $-15°C$ to °F,

$$\frac{9 \times -15}{5} + 32 = 5°F$$

2. To convert degrees Fahrenheit to degrees Celsius (C, centigrade),

$$\frac{5 \times (°F - 32)}{9} = °C, \text{ or } °C = 0.556(°F - 32)$$

To convert 150.8°F to °C,

$$\frac{5 \times (150.8 - 32)}{9} = 66°C$$

To convert $-20°F$ to °C,

$$\frac{5 \times (-20 - 32)}{9} = -29°C$$

F. COMPUTATION OF CENTRIFUGAL FORCE

1. Method 1 [54]

Centrifugal force $= 1118 \times$ radius (in cm) $\times$ RPM$^2 \times 10^{-8}$ or

$$= \frac{\text{Speed}^2 \; R}{89,500}$$

You are using a 200-ml centrifuge bottle in a free-swinging rotor head of 7-in. radius in an International Centrifuge, Type 1, Model SBV, at 700 RPM. What is the centrifugal force?

$$1118 \times 17.5 \text{ cm} \times 490,000 \times 10^{-8} = 95.87 \text{ g}$$

2. Method 2 [11]

Calculations of relative centrifugal force (g) from the radius and the RPM are simplified by the use of the accompanying diagram (Fig. A-3).

Stretch a ruler across the chart to intercept the scales at any two variables at selected values. The ruler will cross the third scale at a value

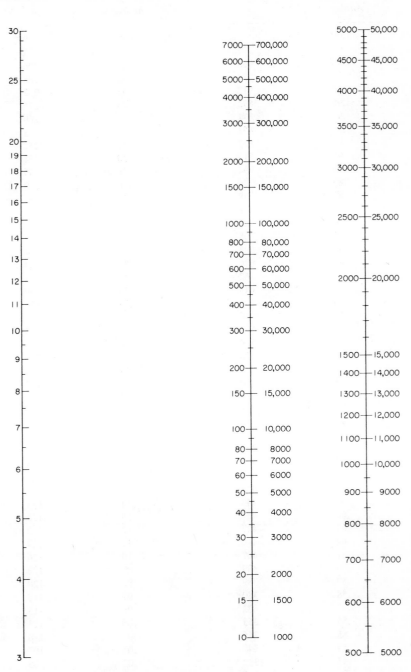

Fig. A-3. Centrifugal force nomogram.

G. LOG-ANTILOG TABLE

Antilog of	is	Antilog of	is
1.00	1	.49	309
.99	977	.48	302
.98	955	.47	296
.97	934	.46	288
.96	912	.45	282
.95	892	.44	276
.94	871	.43	269
.93	851	.42	263
.92	832	.41	257
.91	813	.40	251
.90	794	.39	245
.89	776	.38	240
.88	758	.37	235
.87	741	.36	229
.86	725	.35	224
.85	708	.34	219
.84	693	.33	214
.83	676	.32	209
.82	662	.31	204
.81	645	.30	199
.80	631	.29	195
.79	616	.28	191
.78	602	.27	187
.77	589	.26	182
.76	576	.25	178
.75	563	.24	174
.74	550	.23	170
.73	536	.22	166
.72	525	.21	162
.71	513	.20	159
.70	502	.19	155
.69	490	.18	152
.68	479	.17	148
.67	466	.16	145
.66	457	.15	142
.65	447	.14	138
.64	437	.13	135
.63	427	.12	132
.62	416	.11	129
.61	407	.10	126
.60	399	.09	123
.59	389	.08	120
.58	380	.07	118
.57	372	.06	115
.56	363	.05	113
.55	355	.04	110
.54	347	.03	107
.53	339	.02	105
.52	332	.01	103
.51	324	.00	100
.50	316		

which satisfies the equation

$$\text{Centrifugal force (g)} = 1.11 \times 10^{-5}(R) \times (\text{RPM})^2$$

You can calculate values outside the indicated ranges by shifting the decimal point. A change in the radius by a factor of 10 alters the centrifugal force by the same extent. A change in RPM by a factor of 10 affects the centrifugal force by a factor of 100.

B. Sources of Supplies and Equipment

A. DIAGNOSTIC REAGENTS

Antigens, antisera, blood cells, complement, hemolysin, etc.

1. Baltimore Biological Laboratories,
   Division of BioQuest,
   P.O. Box 175,
   Cockeysville, Md. 21030.

2. Certified Blood Donor Service,
   400 Crossways Park Drive,
   Woodbury, N.Y. 11797.

3. Difco Laboratories, Inc.,
   Detroit, Mich. 48201.

4. General Diagnostics Division,
   Warner-Chilcott Laboratories,
   Morris Plains, N.J. 07950.

5. Hyland,
   Division of Travenol Laboratories,
   Inc., Los Angeles, Calif. 90039.

6. Lederle Laboratories,
   Pearl River, N.Y. 10965.

7. Ortho Diagnostics,
   Raritan, N.J. 08869.

8. Pfizer Diagnostics,
   300 West 43rd St.,
   New York, N.Y. 10036.

9. Microbiological Associates,
   4733 Bethesda Ave.,
   Bethesda, Md. 20014.

10. Grand Island Biological Co.,
    3175 Staley Rd.,
    Grand Island, N.Y. 14072.

B. LABORATORY EQUIPMENT AND SUPPLIES

Glassware, magnetic stirrers, Petri dishes, corrugated boxes, cleaning compounds, cotton, etc.

1. Baltimore Biological Laboratories,
   Division of BioQuest,
   P.O. Box 175,
   Cockeysville, Md. 21030.

2. Bellco Glass, Inc.,
   Vineland, N.J. 08360.

3. Corning Glass Works,
   Laboratory Products Division,
   Corning, N.Y. 14830.

4. Falcon Plastics,
   Division of BioQuest,
   5500 W. 83rd St.,
   Los Angeles, Calif. 90045.

5. Lab-Line Instruments, Inc.,
   Melrose Park, Ill. 60160.

6. Fisher Scientific Co.,
   461 Riverside Ave.,
   Medford, Mass. 02155.

7. Millipore Corporation,
   Bedford, Mass. 01730.

8. Precision Scientific Co.,
   3737 W. Cortland St.,
   Chicago, Ill. 60646.

9. Professional Tape Co., Inc.,
   355 E. Burlington Rd.,
   Riverside, Ill. 60546.

10. Connecticut Chemical and
    Disinfectant Co., 683 Chapel St.,
    New Haven, Conn. 06510.
11. Johnson and Johnson,
    4949 W. 65th St.,
    Chicago, Ill. 60638.
12. Connecticut Container Corp.,
    State-Sackett Point Rd.,
    New Haven, Conn. 06510.
13. Kendall Co.,
    Fiber Products Division,
    Walpole, Mass. 02081.
14. Nathan Schwarz and Sons, Inc.,
    630 Water St.,
    Bridgeport, Conn. 06603.

15. Republic Seitz Filter Corp.,
    P.O. Box 229,
    Milldale, Conn. 06467.
16. West Chemical Co.,
    New Haven, Conn. 06510.
17. Southern New England
    Ultraviolet Co., Newfield St.,
    Middletown, Conn. 06457.
18. Scientific Glass Apparatus Co., Inc.,
    Bloomfield, N.J. 07003.
19. Vollrath Co.,
    1236 N. 18th St.,
    Sheboygan, Wis. 53081.
20. Becton, Dickinson and Co.,
    Rutherford, N.J. 07070.

C. COMMERCIALLY PREPARED MEDIA

1. Baltimore Biological Laboratories,
   Division of BioQuest,
   P.O. Box 175,
   Cockeysville, Md. 21030.
2. Difco Laboratories, Inc.,
   Detroit, Mich. 48201.
3. Microbiological Associates,
   4733 Bethesda Ave.,
   Bethesda, Md. 20014.

4. Grand Island Biological Co.,
   3175 Staley Rd.,
   Grand Island, N.Y. 14072.
5. Fisher Scientific Co.,
   461 Riverside Ave.,
   Medford, Mass. 02155.
6. Flow Laboratories, Inc.,
   12601 Twinbrook Parkway,
   Rockville, Md. 20852.

D. VIRUS STRAINS AND CELL CULTURE LINES

1. American Type Culture Collection,
   12301 Parklawn Drive,
   Rockville, Md. 20852.
2. Difco Laboratories,
   Detroit, Mich. 48201.
3. Presque Isle Microbiologicals,
   P.O. Box 8007,
   Presque Isle, Pa. 16505.

4. Microbiological Associates,
   Bethesda, Md. 20014.
5. Grand Island Biological Co.,
   3175 Staley Rd.,
   Grand Island, N.Y. 14072.

Bibliography

1. Andrewes, C. and D. Horstmann, 1949. "The susceptibility of viruses to ethyl ether," *J. Gen. Microbiol.*, **3**: 290–297.

2. Baltimore, D., H. Eggers, R. Franklin, and I. Tamm, 1963. "Poliovirus-induced RNA polymerase and the effects of virus-specific inhibitors on its production," *Proc. Nat. Acad. Sci., U.S.*, **49**: 843–849.

3. Baron, S., J. S. Porterfield, and A. Isaacs, 1961. "The influence of oxygenation on virus growth. I. Effect on plaque formation by different viruses," *Virology*, **14**: 444–449.

4. Calbiochem., 1967. "Lactalbumin medium containing HEPES buffer instead of phosphate," *Biologics*, **51**: 4.

5. Clarke, D. H. and J. Casals, 1958. "Techniques for hemagglutination and hemagglutination-inhibition with arthropod-borne viruses," *Amer. Jour. Trop. Med. Hyg.*, **7**: 561–573.

6. Cohen, S. S., J. G. Flaks, H. D. Barner, M. R. Loeb, and J. Lichtenstein, 1958. "The mode of action of 5-fluorouracil and its derivatives," *Proc. Nat. Acad. Sci., U.S.*, **44**: 1004–1012.

7. Cooper, P. D., 1955. "A method for producing plaques in agar suspensions of animal cells," *Virology*, **1**: 397–401.

8. Cooper, P. D., 1961. "An improved agar cell-suspension plaque assay for Poliovirus: some factors affecting efficiency of plating," *Virology*, **13**: 153–157.

9. DeLamater, E. D., 1948. "Basic fuchsin as a nuclear stain," *Stain Technol.*, **23**: 161–176.

10. Delbrück, M., 1940. "The growth of bacteriophage and lysis of the host," *Jour. Gen. Physiol.*, **23**: 643–660.

11. Dole, V. P. and G. C. Cotzias, 1951. "A nomogram for the calculation of relative centrifugal force," *Science*, **113**: 552.

12. Drescher, J., A. V. Hennessy, and F. M. Davenport, 1962. "Photometric methods for the measurement of hemagglutinating viruses and antibody," *Jour. Immunol.*, **89**: 794–814.

13. Dulbecco, R. and M. Vogt, 1954. "Plaque formation and isolation of pure lines of Poliomyelitis viruses." *Jour. Exptl. Med.*, **99**: 167–182.

14. Earle, W. R., 1943. "Production of malignancy in vitro. IV. The mouse fibroblast culture and changes seen in the living cells," *Jour. Nat. Cancer Inst.*, **4**: 165–212.

15. Ellis, E. L. and M. Delbrück, 1939. "The growth of bacteriophage," *Jour. Gen. Physiol.*, **22**: 365–384.

16. Enders, J. F., T. H. Weller, and F. C. Robbins, 1949. "Cultivation of the

Lansing strain of Poliomyelitis virus in cultures of various human embryonic tissues," *Science*, **109**: 85–87.

17. Feldman, H. and S. Wang, 1961. "Sensitivity of various viruses to chloroform," *Proc. Soc. Exptl. Biol. Med.*, **106**: 736–738.

18. Feulgen, R. and H. Rossenbeck, 1924. "Mikroskopisch-chemischer Nachweis einer Nucleinsäure vom Typus der Thymonucleinsäure und die darauf beruhende elektive Farbung von Zellkernen in mikroskopischen Präparaten," *Zeit. Physiol. Chem.*, **135**: 203–248.

19. Goodpasture, E. W., 1938. "Some uses of the chick embryo for the study of infection and immunity," *Amer. Jour. Hyg.*, **28**: 111–129.

20. Green, R. and E. Opton, 1959. "Photosensitization of tissue culture cells and its effect on viral plaque formation," *Proc. Soc. Exptl. Biol. Med.*, **102**: 519–521.

21. Greiff, D. and W. Rightsel, 1967. "Stabilities of suspensions of viruses after freezing or drying by vacuum sublimation and storage," *Cryobiology*, **3**: 432–444.

22. Hanks, J. H., 1948. "Calcification of cell cultures in the presence of embryo juice and mammalian sera," *Proc. Soc. Exptl. Biol. Med.*, **71**: 328–334.

23. Hanks, J. H. and R. E. Wallace, 1949. "Relation of oxygen and temperature in the preservation of tissues by refrigeration," *Proc. Soc. Exptl. Biol. Med.*, **71**: 196–200.

24. Hotchin, J., 1955. "Use of methyl cellulose gel as a substitute for agar in tissue-culture overlays," *Nature*, London, **175**: 352.

25. Hsiung, G. D. and J. L. Melnick, 1958. "Orphan viruses of man and animals," *Ann. N.Y. Acad. Sci.*, **70**: 342–361.

26. Kärber, G., 1931. "Beitrag zur kollektiven Behandlung pharmakologischer Reihenversuche," *Arch. Exptl. Path. Pharmakol.*, **162**: 480–483.

27. Karstadt, L., J. Spalatin, and R. P. Hanson, 1957. "Application of the paper disc technique to the collection of whole blood and serum samples in studies on Eastern Equine Encephalomyelitis," *Jour. Infec. Diseases*, **101**: 295–299.

28. Ketler, A., V. Hamparian, and M. Hilleman, 1962. "Characterization and classification of ECHO 28-rhinovirus-coryzavirus agents," *Proc. Soc. Exptl. Biol. Med.*, **110**: 821–831.

29. Ledinko, N., 1955. "Production of plaques with Influenza viruses," *Nature*, London, **175**: 999–1000.

30. Liebhaber, H. and K. Takemoto, 1961. "Alteration of plaque morphology of EMC virus with polycations," *Virology*, **14**: 502–504.

31. McManus, J. F. A. and R. W. Mowry, 1960. *Staining Methods: Histologic and Histochemical*, New York: Harper & Row, Pub., Inc., p. 388.

32. McManus, J. F. A. and R. W. Mowry, 1960. *Staining Methods: Histologic and Histochemical*. New York: Harper & Row, Pub., Inc., p. 389.

33. Mayor, H. D., 1961. "Cytochemical and fluorescent antibody studies on the growth of poliovirus in tissue culture," *Texas Repts. Biol. Med.*, **19**: 106–122.

34. Mayor, H. D., 1964. "A labile intranuclear RNA associated with the development of adenoviruses," *Jour. Exptl. Med.*, **119**: 433–441.

35. Melnick, J. L., 1955. "Tissue culture techniques and their application to original isolation, growth, and assay of poliomyelitis and orphan viruses," *Ann. N.Y. Acad. Sci.*, **61**: 754–773.

36. Pledger, R., 1960. "Effect of neutral red on plaque formation by Foot-and-Mouth Disease virus," *Virology*, **10**: 50–56.

37. Prusoff, W. H., Y. S. Bakhle, and L. Sekelv, 1965. "Cellular and antiviral effects of halogenated deoxyribonucleosides," *Ann. N.Y. Acad. Sci.*, **130**: 135–150.

38. Rappaport, C., 1956. "Trypsinization of monkey-kidney tissue: an automatic method for the preparation of cell suspensions," *Bull. Wld. Hlth. Org.*, **14**: 147–166.

39. Reed, L. J. and H. Muench, 1938. "A simple method of estimating fifty per cent endpoints," *Amer. Jour. Hyg.*, **27**: 493–497.

40. Reich, E., R. M. Franklin, A. J. Shatkin, and E. L. Tatum, 1961. "Effect of actinomycin-D on cellular nucleic acid synthesis and virus production," *Science*, **134**: 556–557.

41. Rightsel, W. and Greiff, D., 1967. "Freezing and freeze-drying of viruses," *Cryobiology*, **3**: 423–431.

42. Rous, P. and F. S. Jones, 1916. "A method of obtaining suspensions of living cells from the fixed tissues, and for the plating out of individual cells," *Jour. Exptl. Med.*, **23**: 549–562.

43. Rous, P. and J. B. Murphy, 1911. "Tumor implantation in developing embryos," *Jour. Amer. Med. Assn.*, **56**: 741–742.

44. Sabin, A. B. and P. K. Olitsky, 1936. "Cultivation of Poliomyelitis virus in utero in human embryonic nervous tissue," *Proc. Soc. Exptl. Biol. Med.*, **34**: 357–359.

45. Salk, J., 1944. "A simplified procedure for titrating hemagglutinating capacity of influenza-virus and the corresponding antibody," *Jour. Immunol.*, **49**: 87–98.

46. Salk, J., U. Krech, J. Youngner, B. Bennett, L. Lewis, and P. Bazeley, 1954. "Formaldehyde treatment and safety testing of experimental poliomyelitis vaccines," *Amer. Jour. Public Health*, **44**: 563–570.

47. Salzman, N. P., A. J. Shatkin, and E. D. Sebring, 1964. "The synthesis of a DNA-like RNA in the cytoplasm of HeLa cells infected with Vaccinia virus," *Jour. Mol. Biol.*, **8**: 405–416.

48. Schmidt, N., V. Fox, and E. Lennette, 1962. "Studies on the hemagglutination of COE (Coxsackie A 21) virus," *Jour. Immunol.*, **89**: 672–683.

49. Schmidt, N. J., R. W. Guenther, and E. H. Lennette, 1961. "Typing of ECHO virus isolates by immune serum pools; the intersecting serum scheme," *Jour. Immunol.*, **87**: 623–626.

50. Seligman, S. J. and M. R. Mickey, 1964. "Estimation of the number of infectious bacterial or viral particles by the dilution method," *Jour. Bacteriol.*, **88**: 31–36.

51. Shelokov, A., J. E. Vogel, and L. Chi, 1958. "Hemadsorption (adsorption-hemagglutination) test for viral agents in tissue culture with special reference to Influenza," *Proc. Soc. Exptl. Biol. Med.*, **97**: 802–809.

52. Tamm, I. and H. Eggers, 1963. "Specific inhibition of replication of animal viruses," *Science*, **142**: 24–33.

53. Tytell, A. A. and R. E. Neuman, 1963. "A medium free of agar, serum and peptone for plaque assay of Herpes simplex virus," *Proc. Soc. Exptl. Biol. Med.*, **113**: 343–346.

54. Umbreit, W. W., R. H. Burris, and J. F. Stauffer, *Manometric Techniques*, 4th ed., Chicago, Ill.: Burgess Publishing Co., p. 180.

55. Ver, B. A., J. L. Melnick, and C. Wallis, 1968. "Efficient filtration and sizing of viruses with membrane filters," *Jour. Virology*, **2**: 21–25.

56. Vogel, J. and A. Shelokov, 1957. "Adsorption-hemagglutination test for Influenza virus in monkey kidney tissue culture," *Science*, **126**: 358–359.

57. Wallis, C. and J. Melnick, 1961. "Stabilization of poliovirus by cations," *Texas Repts. Biol. Med.*, **19**: 683–700.

58. Wallis, C., J. L. Melnick, and M. Bianchi, 1962. "Factors influencing enterovirus and reovirus growth and plaque formation," *Texas Repts. Biol. Med.*, **20**: 693–702.

59. Wallis, C., C. Yang, and J. L. Melnick, 1962. "Effect of cations on thermal inactivation of Vaccinia, Herpes simplex, and adenoviruses," *Jour. Immunol.*, **89**: 41–46.

60. Weller, T. H., J. F. Enders, F. C. Robbins, and M. B. Stoddart, 1952. "Studies on the cultivation of poliomyelitis viruses in tissue culture. I. The propagation of poliomyelitis virus in suspended cell cultures of various human tissues," *Jour. Immunol.*, **69**: 645–671.

61. Youngner, J. S., 1954. "Monolayer tissue cultures; I. Preparation and standardization of suspensions of trypsin-dispersed monkey kidney cells," *Proc. Soc. Exptl. Biol. Med.*, **85**: 202–205.

INDEX

A

Abdominal fluid, 8
Acid-citrate-dextrose, 232
  procedure, 159-163
  stain preparation, 240
Actinomycin D
  sensitivity of viruses to, 151-152
Agar
  bottom layer, 229
  nutrient, 228
  plaque suspension, 229
  top layer, 228
Animals
  experimental, general guidepoints
    in handling, 68-69
  harvest and repassage of tissues from,
    1-15, 75-76
  inoculations
    effects of, 69
  necropsy, guidepoints in handling, 3,
    75
  recording appropriate information on

Animals *(cont.)*
    experimental animals, 75, 86
  restraint methods, 69
  routes of inoculation for, 69-75
    ingestion, 70
    inhalation, 70
    intracutaneous, 74
    intramuscular, 73
    intraperitoneal, 74
    intravenous, 73
    subcutaneous, 74
  specimen collection from, 1-15, 75-76
Anhydrous-hydrous chemical conversion
    calculations, 267
Antibiotic preparations, 235
Avian embryo fibroblast, 41-42

B

Bacteriophage, 164-179
  broth method of propagation, 171-172

281

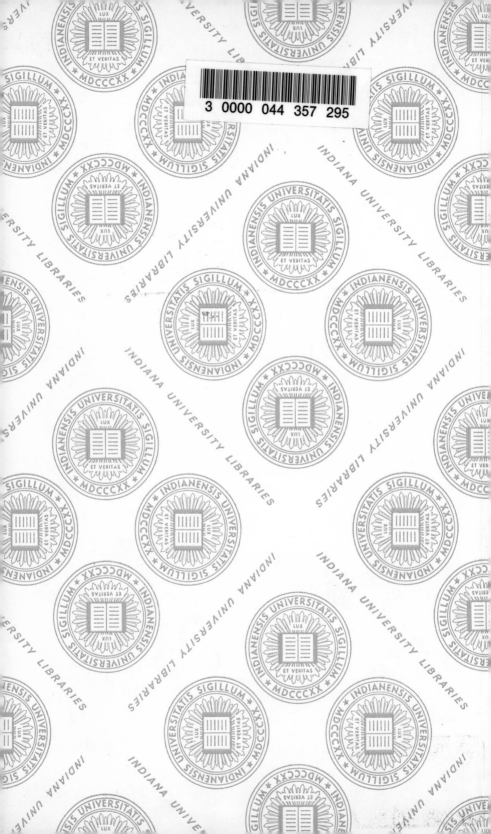